SpringerBriefs in Molecular Science

Green Chemistry for Sustainability

Series editor

Sanjay K. Sharma, Jaipur, India

More information about this series at http://www.springer.com/series/10045

Andrew P. Dicks · Andrei Hent

Green Chemistry Metrics

A Guide to Determining and Evaluating Process Greenness

 Springer

Andrew P. Dicks
Andrei Hent
Department of Chemistry
University of Toronto
Toronto, ON
Canada

ISSN 2212-9898
ISBN 978-3-319-10499-7   ISBN 978-3-319-10500-0   (eBook)
DOI 10.1007/978-3-319-10500-0

Library of Congress Control Number: 2014950410

Springer Cham Heidelberg New York Dordrecht London

Printed on acid-free paper

Springer is part of Springer Science+Business Media (www.springer.com)

Preface

If the idols of scientists were piled on top of one another
in the manner of a totem pole the topmost
would be a grinning fetish called Measurement

Anthony Standen, *Science is a Sacred Cow*

Green chemistry is currently a field of great interest to many scientists, but is sometimes perceived as one rooted in descriptive language built around reducing waste and material recycling. This book seeks to outline a quantitative approach to green chemistry, at a level easily understood by upper-level undergraduates. It is written for students, and for university/college instructors seeking to "green" their courses by framing discussion around four principal metrics. In doing this, one will *REAP* the rewards of learning about *R*eaction mass efficiency, *E* factor, *A*tom economy, and *P*rocess mass intensity as measurement tools toward a more sustainable future.

These four metrics might be discussed within, for example, an organic synthesis course for chemistry students, or as part of a "stand-alone" green chemistry offering (the type of which is becoming more popular in education around the world). In addition, time is taken to cover more qualitative methods toward green chemistry assessment, and the notion of life cycle analysis. All of these concepts are presented in the context of cutting-edge academic and industrial case studies that showcase the state of the art. It is sincerely hoped that learning about these topics will empower students to make informed decisions regarding greener technologies in their future work.

Toronto, July 2014

Andrew P. Dicks
Andrei Hent

Contents

Chapter 1
Green Chemistry and Associated Metrics

Abstract This chapter provides an overview of green metrics and their historical role in promoting the development of green chemistry. Starting with the history of the field, the Twelve Principles of Green Chemistry are introduced and discussed in conjunction with a "green-by-design" approach recently applied to the synthesis of Lipitor®. Various perspectives on synthetic efficiency are briefly outlined with reference to atom economy and E factor. These ideas are further explored in the context of three industrial processes which have received Presidential Green Chemistry Challenge Awards. The synthesis of ibuprofen is examined from the point of view of intrinsic efficiency. Using the BHC process as an example, several benefits associated with the use of catalysis are discussed, with an emphasis placed on designing atom-efficient reactions. A global perspective centered around the production of chemical waste is also outlined with reference to Merck's commercial synthesis of Januvia®, a medication for the treatment of type II diabetes. Finally, Pfizer's new sertraline process is used to describe ways of improving both quantitative as well as qualitative aspects of an industrial synthesis. The chapter concludes with a brief outline of the future directions of green metrics.

Keywords Principles of green chemistry · Synthetic efficiency · Presidential green chemistry challenge awards · Lipitor · Ibuprofen · Januvia · Sertraline · Atom economy · E factor

1.1 Green Chemistry

One of the most influential ideas in the physical sciences today is the notion of green chemistry. The concept originated in 1991 when Paul Anastas of the United States Environmental Protection Agency (EPA) helped launch the Green Chemistry Program [1]. The aim of this initiative was "to promote the research, development, and implementation of innovative chemical technologies that accomplish pollution prevention in both a scientifically sound and cost-effective manner" [1]. This

© The Author(s) 2015
A.P. Dicks and A. Hent, *Green Chemistry Metrics*,
SpringerBriefs in Green Chemistry for Sustainability,
DOI 10.1007/978-3-319-10500-0_1

environmentally-conscious approach was also fuelled by the publication of two major articles on green metrics [2, 3] which contributed to the development of the Twelve Principles of Green Chemistry in 1998 [4]. Green chemistry has since become a powerful idea appearing in thousands of research articles, numerous books on fundamental and current research [5, 7], and many pedagogical works that aim to incorporate the subject into the university curriculum [8–10]. An essential aspect of the movement pertains to the development of appropriate metrics. The purpose of these methods is to change the dominant view that process efficiency can be accurately described using a synthetic product yield. By adopting larger perspectives on synthetic efficiency, scientists are empowered to improve the quality and sustainability of the processes they are asked to design.

1.1.1 The Twelve Principles of Green Chemistry

In 1991, under the EPA's Green Chemistry Program Fact Sheet, the goal of the discipline was identified as "the design of chemical products and processes that reduce or eliminate the use or generation of hazardous substances" [1]. In 1998, Anastas and Warner formulated twelve fundamental principles to help others achieve this goal [4]:

1. **Prevention of waste**. The first and most important principle states that waste prevention is better than waste treatment.
2. **Atom economy**. Chemists are encouraged to design synthetic methods which maximize the incorporation of starting materials into the final product.
3. **Safer synthesis**. It is important, wherever practicable, to design chemical methods which require and generate substances with little or no toxicity to people and the environment.
4. **Safer chemicals**. Chemical products should be designed to achieve a desired function with minimal toxicity.
5. **Safer solvents and auxiliaries**. Auxiliary substances such as solvents and separation agents should be eliminated whenever possible and made innocuous when used.
6. **Energy efficiency**. It is important to identify and minimize the environmental and economic impacts associated with energy use in chemical synthesis. Methods conducted at ambient temperature and pressure should be adopted whenever possible.
7. **Renewable feedstocks**. Starting materials originating from renewable feedstocks should be used whenever technically and economically practicable.
8. **Fewer derivatives**. The use of blocking groups, protection/deprotection, and temporary modification of physical/chemical processes should be minimized or excluded for waste reduction purposes.

9. **Catalysis.** When compared to stoichiometric reagents, catalytic reagents that are engineered for selectivity are more efficient and generally produce less waste.

10. **Design for degradation.** It is important to design chemical products which break down into innocuous degradation materials at the end of their function and which do not persist in the environment.

11. **Real-time analysis.** It is important to develop and adopt analytical methods that provide real-time, in-process monitoring and control prior to the formation of hazardous substances.

12. **Safer chemistry for accident prevention.** The potential for chemical accidents such as releases, explosions, and fires should be minimized by choosing inherently safer substances.

For a more complete presentation of the Twelve Principles the reader is referred to the revised paperback version of a 1998 publication by Anastas [11]. It is also useful to recognize that the first two principles of green chemistry describe the E factor [3] and atom economy metrics [2] (Chaps. 3 and 2 respectively). The remaining ten principles serve to guide environmentally-conscious design within chemical synthesis. Such an approach, coined "green-by-design", was recently applied in planning a greener synthesis of atorvastatin calcium (Fig. 1.1), the active ingredient in the blockbuster drug Lipitor® [12].

Lipitor has been marketed by Pfizer as a cholesterol-lowering medication since 1996, and is the first ever drug to reach annual sales of $10 billion [12]. Today, even with its patent expired, Lipitor continues to generate annual sales of $3 billion in the United States [13]. Sales of this magnitude provide a great incentive for designing a more efficient process. With the aid of molecular biology technologies, Ma et al. engineered an enzyme for a biocatalytic process which helps produce atorvastatin calcium more safely and efficiently [12]. The authors described the green features of the new process in the context of each of the Twelve Principles, highlighting their power in directing effort and innovation toward the development of greener synthetic methods.

The philosophy was further extended in 2001 when Winterton devised twelve additional principles [14], and in 2003 when Anastas published the Twelve Principles of Green Engineering [15]. In subsequent years, Poliakoff et al. proposed the useful mnemonics PRODUCTIVELY and IMPROVEMENTS to account for the

Fig. 1.1 Structure of atorvastatin calcium

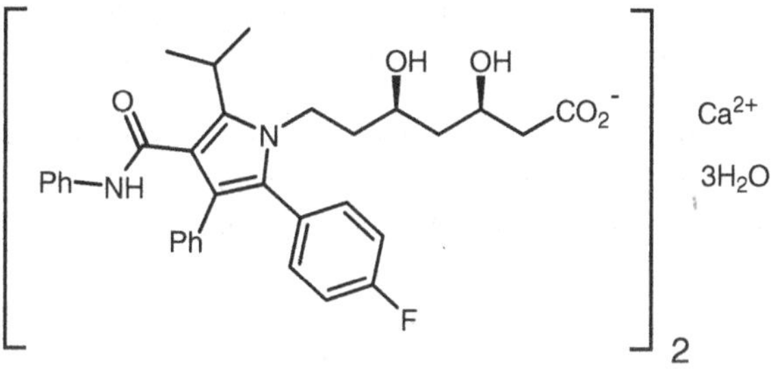

Twelve Principles of Green Chemistry and Green Engineering respectively [16, 17]. Since then, numerous reviews have outlined the state and future direction of green chemistry from the point of view of the Twelve Principles [18–22] and literature surveys in general [7, 8, 23, 24].

1.1.2 Synthetic Efficiency and Overall Process Quality

Green chemistry principles have enabled chemists to propose more integrated definitions of the subject. One such definition states that "green chemistry efficiently utilizes (preferably renewable) raw materials, eliminates waste and avoids the use of toxic and/or hazardous reagents and solvents in the manufacture and application of chemical products" [25]. Although this definition is clear, it does not enable an objective comparison of two different procedures. In practice, this raises certain issues since real-world syntheses generally exhibit various "shades of green" which are often overlooked in research articles. Although academic journals are starting to change their submission policies, articles which lack an assessment of asserted greenness are still common. It is often under-appreciated that whenever a method is claimed to be green, the author should be responsible for disclosing sufficient experimental details to enable an assessment of the entire process, including features which may contradict the principles of green chemistry. Furthermore, even though compromises may be unavoidable, quantitative environmental assessments can help drive the improvement of a particular synthesis.

The simplest way to begin an environmental assessment is by considering waste. Within the totality of a chemical process, waste can assume both material and non-material forms. The first two principles of green chemistry provide a material interpretation of waste. The remaining principles attempt to set standards for assessing qualitative aspects of waste (based on environmental implications) which are not easily accounted for quantitatively. Consequently, the purpose of a green metrics analysis is to determine the extent to which a chemical synthesis achieves the definition of green chemistry both quantitatively and qualitatively. To better understand this purpose, it is useful to consider a chemical transformation from several different perspectives.

A straightforward approach reflects on the intrinsic efficiency of a chemical reaction. At this level, one studies the fate of reactant atoms in terms of whether or not they are incorporated into the desired product. A high intrinsic efficiency generally favours addition and rearrangement reactions (sometimes in a catalytic environment). Catalysis is emphasized because reagents used in stoichiometric quantities are generally factored into the calculation of atom economy (Chap. 2) [2, 26]. It is also important to note that intrinsic efficiency necessarily ignores other reaction components such as solvents and catalysts. In the case of solvents, previous work has concluded that they often represent a significant portion of the process mass balance [27]. For this reason, it is also important to consider a chemical synthesis from a global perspective.

Sheldon approached this problem by introducing the E factor in 1992 (Chap. 3), a metric which measures the mass of generated waste per unit mass of final product formed [3]. It is useful to note that the E factor mathematically incorporates all the information provided by atom economy into a larger, more encompassing mass-based metric [28]. Some of this information is not usually reflected by the atom economy metric since the E factor typically measures large solvent masses (destined for waste). Although atom economy considerations may motivate research into more atom-efficient reactions, the E factor advances a global approach to waste reduction at all stages of experimental design. Nevertheless, adopting both intrinsic and global perspectives enables a better understanding of process features and dynamics.

Lastly, considering the nature of the chemicals used as well as their source and ultimate destination can help improve the safety and environmental profile of a process. Since scientific consensus regarding these parameters has not yet been established, qualitative assessments of a process have often been more difficult to perform. To shed some light on these topics, Chaps. 4 and 5 have been devoted to selected qualitative metrics and life cycle analysis tools respectively. In the following sections, perspectives on process greenness mentioned thus far are explored by reference to some well-established industrial achievements in green chemistry.

1.2 Some Award-Winning Green Processes

1.2.1 The Presidential Green Chemistry Challenge Awards

One of the most important contributions of the EPA Green Chemistry Program was the establishment of the Presidential Green Chemistry Challenge Awards in 1996. With the aim of promoting "the environmental and economic benefits of developing and using novel green chemistry", winning technologies recognized thus far have collectively produced remarkable results [29]. According to EPA statistics, the 93 winning technologies have contributed to an annual reduction of 826 million pounds of hazardous chemicals and solvents, 21 billion gallons of water, and 7.8 billion pounds of CO_2 equivalents that would otherwise have been released into the atmosphere [29]. This program has also been described as a useful approach for introducing students to green chemistry [30]. In the next sections, three highly publicized Presidential Green Chemistry Challenge Award processes are presented in the context of synthetic efficiency and process quality. These achievements will highlight the power of applying green metrics to the study and development of green chemistry. It is also worth noting that the redesigned synthesis of atorvastatin calcium outlined briefly in Sect. 1.1.1 was a Presidential Green Chemistry Challenge Award winner in 2006 [31].

1.2.2 BHC Ibuprofen Synthesis: A Perspective on Intrinsic Efficiency

When deciding on ways to improve a chemical process, one of the first essentials to consider is the synthetic scheme involved. Firstly, synthetic schemes inform chemists about which chemicals need to be purchased and in what quantities. Secondly, they provide estimates regarding how much final product can be generated given the chosen reactions. Finally, the atoms which are incorporated into the final product are differentiated from those which are ultimately destined for waste. This information forms the basis of the intrinsic efficiency of the chemistry involved in a process. Although intrinsic efficiency does not necessarily determine product yield, it impacts waste in the form of by-product formation. Processes where the majority of atoms do not end up in the final product have a poor intrinsic efficiency, and the greater waste produced often accompanies negative environmental and economic consequences. Designing atom-efficient processes can therefore be extremely beneficial, especially for the manufacturers of high-volume chemical products. These benefits became apparent in 1991 when the Boots Hoechst Celanese (BHC) Company patented a novel process for the manufacture of ibuprofen (Fig. 1.2) [32].

Ibuprofen is an over-the-counter medication prescribed for the treatment of mild to moderate pain. The drug was originally discovered at the Boots Company in 1961 [33]. Ibuprofen functions by inhibiting cyclooxygenase (COX), an enzyme which participates in the arachidonic acid cascade towards prostaglandin synthesis. Prostaglandins are important mediators of inflammation and pain in the human body [34]. After being approved for over-the-counter use, annual ibuprofen prescriptions have grown to 20 million encompassing such brands as Advil, Motrin, Nuprin, Rufen and Trendar [35]. Until 1992, ibuprofen manufacturing consisted of a wasteful and intrinsically inefficient six-step synthesis (Boots process, Scheme 1.1) [36, 37].

The Boots synthesis begins with a Friedel-Crafts acylation of isobutylbenzene **1** using acetic anhydride as the acylating agent in the presence of aluminum trichloride. The resulting isobutylacetophenone **2** then undergoes a Darzens condensation reaction [38] with an α-chloroenolate (formed by reacting an α-chloroester with sodium ethoxide) to produce epoxide **3**. When treated with aqueous acid, **3** undergoes decarboxylation and rearrangement to produce aldehyde **4** which is converted into the oxime **5** using hydroxylamine. The oxime product then

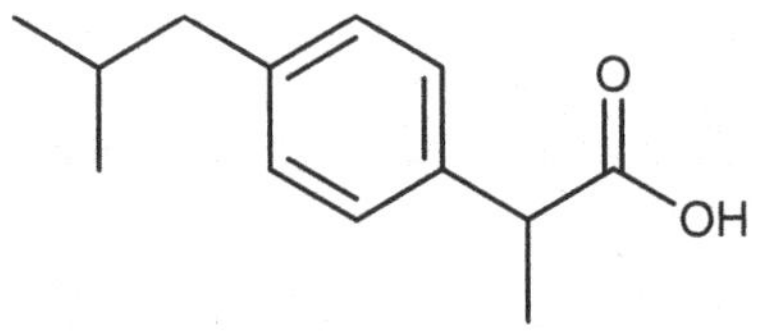

Fig. 1.2 Structure of ibuprofen

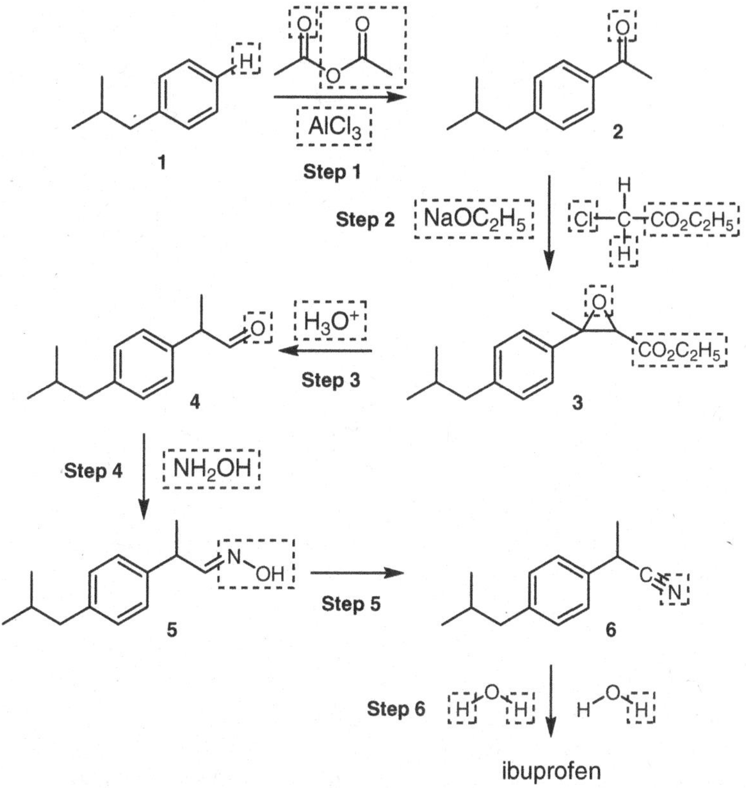

Scheme 1.1 Traditional Boots Company synthesis of ibuprofen (atoms which are not part of the final product are displayed in *broken line boxes*)

spontaneously loses water to form nitrile **6** which produces ibuprofen upon aqueous hydrolysis. Scheme 1.1 illustrates the atoms which do not end up in the final product. One can readily observe that the majority of reagents are not utilized efficiently. Using the atom economy metric (Chap. 2) one can calculate an atom economy of 40 % for the Boots ibuprofen synthesis [36], a value which implies that 60 % of the atomic weight of all reactants involved in the process is ultimately wasted.

As the cost of handling waste steadily increased over time, researchers began to realize that a more efficient synthesis of ibuprofen had to be developed. This approach was realized in 1992 when the newly-formed BHC Company established a highly efficient three-step catalytic synthesis (Scheme 1.2) [32]. In this procedure, isobutylbenzene **1** undergoes a novel Friedel-Crafts acylation in the presence of catalytic amounts of anhydrous hydrogen fluoride, which also functions as the solvent (thus eliminating the need for a co-solvent).

The resulting isobutylacetophenone **2** undergoes heterogeneous catalytic hydrogenation over a solid Raney Nickel catalyst to form the secondary alcohol **7**.

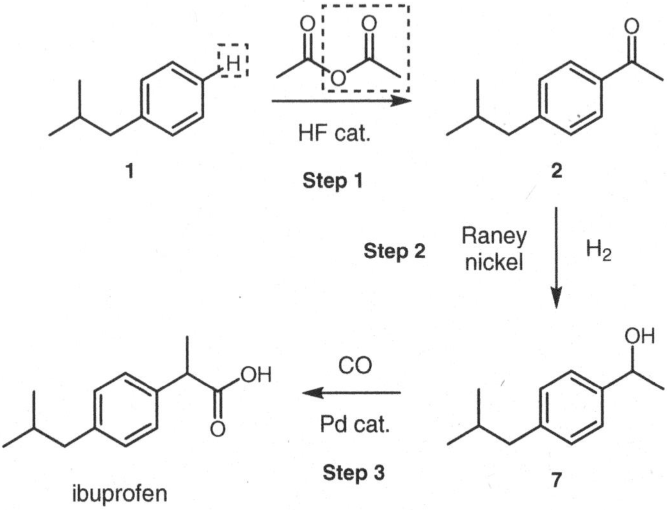

Scheme 1.2 Modern BHC synthesis of ibuprofen (atoms which are not part of the final product are displayed in *broken line boxes*)

The final step constitutes a homogeneous catalysis involving carbonylation of **7** with carbon monoxide in the presence of a soluble palladium catalyst to produce ibuprofen in high yield after vacuum distillation.

The two schemes clearly show that the BHC ibuprofen process is much shorter in length, and significantly more atom-efficient than the Boots method. In terms of atom economy, the BHC process has an atom economy of 77 % [36], a value which rises to 99 % when the recovery of acetic acid (generated in step 1) is considered. Compared to the Boots method which uses many more reagents, the BHC process establishes a significant reduction in waste. This accomplishment led to the BHC process being presented with a Presidential Green Chemistry Challenge Award in 1997 [29]. Some additional features of the BHC synthesis include easier separation procedures as well as the recovery and recycling of all catalysts. Since 1997, the BHC ibuprofen process has appeared in numerous reviews and pedagogical articles as an excellent example that highlights the benefits of catalytic reactions and their ability to improve the intrinsic efficiency of chemical processes [5, 6, 30, 39, 40].

1.2.3 Merck's Synthesis of Januvia: Highlights and Global Efficiency

The foundational principle of green chemistry states that "it is better to prevent waste than to treat or clean up waste after it is formed" [11]. In the previous section, waste prevention was demonstrated on the basis of intrinsic efficiency, an approach

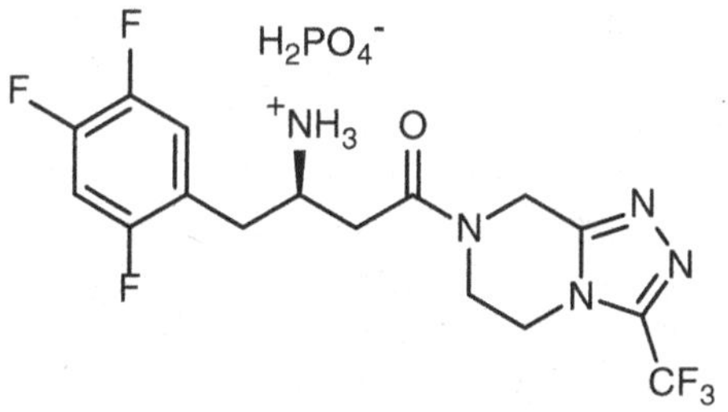

Fig. 1.3 Structure of the monohydrate phosphate salt of sitagliptin

which promotes the design of atom-efficient reactions where catalytic reagents are generally superior to reagents used in stoichiometric quantities. The purpose of this section is to extend the notion of efficiency to encompass the quantities of all materials involved in a particular synthesis, including the separation, isolation and purification steps. To account for this global perspective, the environmental (E) factor metric (Chap. 3) is employed [3]. The purpose of the E factor is to calculate the mass of waste per unit mass of desired product, where waste consists of everything that is not incorporated into the product [3]. Since intrinsic efficiency ignores reaction components such as solvents and catalysts (including their environmental effects) it is vital to adopt a global perspective on processes, especially in an industrial setting. In addition, the global perspective also complements the industrial goal of lowering production costs. Practically, this can be achieved by designing high yielding atom-efficient processes which have the lowest possible E factors. A well-established example of this approach is the redesigned synthesis of sitagliptin phosphate (Fig. 1.3) [41].

Sitagliptin is the active ingredient in Januvia®, a drug marketed by Merck for the treatment of type II diabetes [42]. Since 2005, the manufacturing process for sitagliptin has undergone three stages of development [43–45]. With each new stage, the process has become increasingly more efficient in the context of global waste reduction. For detailed comparisons of the three routes as well as a discussion of the development process, the reader is referred to several excellent reviews [41, 46, 47].

In terms of global efficiency, the initial process developed by Merck (called the β-lactam route) [41] consisted of nine steps with an overall yield of 45 % [46]. This process generated large amounts of waste due to many factors, including the use of a 1-ethyl-3-(3-dimethylaminopropyl)carbodiimide (EDC) coupling and Mitsunobu sequence to convert a hydroxyl group into an amine. It is generally understood that the Mitsunobu reaction has a very poor atom economy ([8], pp 10–12). The expenditure of other high molecular weight reagents used in stoichiometric quantities also contributed to waste production. Given synthetic steps which required reaction volumes of up to 30 L per kg of intermediate formed, the β-lactam route to sitagliptin produced an E factor of 265 (205 kg of organic waste and 60 kg of aqueous waste) [47]. Such a large value makes it economically impractical for a company to bring this process to a commercial scale.

Consequently, researchers at Merck began to work with scientists at Solvias and Codexis (companies specializing in homogeneous catalysis and biocatalysis respectively) in order to develop a more efficient process [41]. The 2006 second-generation synthesis (an asymmetric hydrogenation route) succeeded in lowering the E factor to 67 (65 kg organic waste and 2 kg aqueous waste) [47]. This was achieved by eliminating the need for protecting groups and replacing the costly Mitsunobu sequence with a homogeneous catalytic hydrogenation to transform an enamine group into the necessary chiral amine. To acknowledge this work, Merck was awarded the 2006 Presidential Green Chemistry Challenge Award for Greener Synthetic Pathways [48].

Further, collaboration in 2010 resulted in the third generation synthesis of sitagliptin phosphate (a transaminase biocatalytic route) [46]. Using a newly-engineered transaminase enzyme to circumvent a costly ruthenium based catalytic hydrogenation, the process achieved a 13 % increase in overall yield and a 19 % reduction in overall waste [41]. The biocatalysis afforded milder reaction conditions as well as the ability to conduct the process within multi-purpose reactors thus eliminating the need for costly high-pressure equipment. As a result, this innovation was awarded the 2010 Presidential Green Chemistry Challenge Award for Greener Reaction Conditions [49].

1.2.4 Pfizer's Sertraline Process: A Perspective on Overall Process Quality

Having considered both intrinsic and global efficiency, an emphasis needs to be placed on a more subtle aspect of a chemical synthesis: *the nature of the chemicals and techniques chosen.* Being able to quantify process safety and chemical toxicity using green metrics has proved a very challenging endeavour. Although research focused on identifying [50] and designing [51] safer chemicals continues to occur, the development of a unified green metrics approach has been fraught with subjectivity and lack of scientific consensus (e.g. the EcoScale [52], an approach described in Chap. 4). Nevertheless, progress has been made with regard to use of greener chemicals and techniques in synthesis. A well-documented achievement in this context pertains to the redesigned preparation of sertraline (Fig. 1.4) [53].

The discovery of sertraline (the active ingredient responsible for the antidepressant function of Zoloft®) dates back to the early 1970s [54, 55]. The traditional medicinal chemistry route was developed in 1984 [56] and eventually influenced Pfizer's first commercial process for Zoloft in 1991 (Scheme 1.3) [57]. This synthesis required over 101 L of solvent for every kilogram of sertraline synthesized [57]. Such a large amount of materials was required due to a wasteful imination reaction which occurred in a solvent mixture of toluene, hexanes and tetrahydrofuran (THF). In addition, the corrosive and extremely hazardous titanium

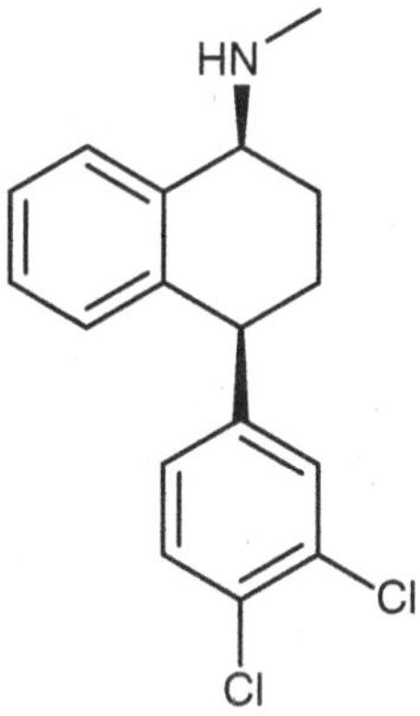

Fig. 1.4 Structure of sertraline

tetrachloride was needed as a dehydrating agent to further shift the equilibrium and drive the reaction to completion.

Repeated solvent switches between THF, ethanol and ethyl acetate during the next three steps also contributed to waste production. Finally, a complicated workup and purification sequence was needed to access the optically active form of sertraline. Overall, the original process consumed 34 L of ethanol, 12 L of hexane, 8 L of toluene, 19 L of THF and 28 L of ethyl acetate per kilogram of sertraline product.

By comparison, Pfizer's 1998 third-generation synthesis (referred to as the combined sertraline process) required only 24 L of solvent per kilogram of ser-traline. This amount consisted of 9 L of ethyl acetate and 15 L of ethanol. What allowed for this significant reduction in solvent use was implementation of ethanol in the first three steps of the process, permitting their consolidation under a one-pot synthesis. Furthermore, use of ethanol eliminated the need for titanium tetrachlo-ride, since the low solubility of the imine product was used to drive the imination reaction to completion. Use of a more selective hydrogenation catalyst further reduced the amount of by-products formed thus significantly simplifying the purification procedure. An improved yield combined with large savings in solvents used, energy expended and raw materials earned this process the 2002 Presidential Green Chemistry Challenge Award for Greener Synthetic Pathways [58].

Furthermore, the redesigned sertraline process illustrates the improvement that is possible with regard to the overall quality of a chemical process. According to Sanofi's recently published solvent selection guide [50], the three solvents elimi-nated under the new sertraline process (toluene, hexane, and THF) are not con-sidered green, and are therefore not recommended for use in synthesis. Conversely, ethanol and ethyl acetate are considered greener solvents which are highly rec-ommended for use in a commercial setting. Since these aspects are not included in a green metrics analysis, it is important to use criteria like industrial solvent selection guides [50, 59, 60] to help evaluate the greenness of a process in terms of its overall quality.

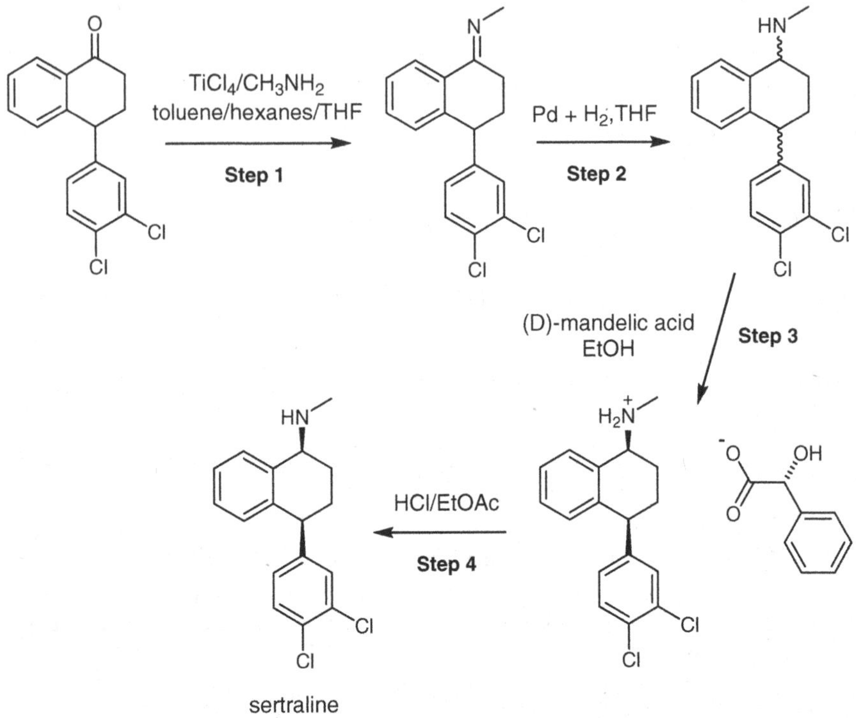

Scheme 1.3 Pfizer's first commercial synthesis of sertraline

1.3 Green Metrics: Overview and the Path Forward

The case studies discussed thus far have illustrated how different approaches can relate key aspects of a chemical process with the principles of green chemistry. Since these methods will be covered in greater detail in the following chapters, an appropriate way to conclude this introduction is to identify and summarize what green metrics actually are and what they do.

Essentially, metrics enable a meaningful assessment of process greenness corresponding with the Twelve Principles of Green Chemistry. In the context of a complete process, it is important to note that no single metric can provide the entire story of synthetic greenness and efficiency. In real-world scenarios, it is often the case that a certain synthesis may encompass various "shades of green". This is why choosing appropriate metrics to identify and describe all features of a transformation is very important. In this context, traditional assessments based solely on the product yield are becoming obsolete. In addition, others have emphasized the need for a more complete disclosure of experimental and methodological details in the literature as to enable readers to properly assess the merits of an improved process [9]. This is especially important with regard to process features that may not be deemed green. In short, a basic understanding of green metrics (including the ways to calculate and interpret the information they supply) is essential to assessing and encouraging progress in green chemistry.'

References

1. Office of pollution prevention and toxics, EPA (2002) Green chemistry program fact sheet. http://nepis.epa.gov/Exe/ZyPDF.cgi/P1004H5E.PDF?Dockey=P1004H5E.PDF. Accessed 28 Apr 2014
2. Trost BM (1991) The atom economy—a search for synthetic efficiency. Science 254:1471–1477. doi:10.1126/science.1962206
3. Sheldon RA (1992) Organic synthesis—past, present and future. Chem Ind 903–906
4. Anastas PT, Warner JC (1998) Green chemistry: theory and practice. Oxford University Press, Oxford 29
5. Lancaster M (2010) Green chemistry: an introductory text, 2nd edn. RSC Paperbacks, Cambridge
6. Sheldon RA, Arends IWCE, Hanefeld U (2007) Green chemistry and catalysis. Wiley-VCH Verlag, Weinheim
7. Anastas P (2010–2012) Handbook of green chemistry vol 1–9. Wiley-VCH Verlag GmbH & Co. KGaA, Weinheim
8. Dicks AP (2012) Green organic chemistry in lecture and laboratory. CRC Press, Taylor and Francis Group, Boca Raton
9. Andraos J (2012) The algebra of organic synthesis: green metrics. CRC Press, Taylor and Francis Group, Boca Raton
10. Dicks AP, Batey RA (2013) ConfChem conference on educating the next generation: green and sustainable chemistry—greening the organic curriculum: development of an undergraduate catalytic chemistry course. J Chem Educ 90:519–520. doi:10.1021/ed2004998
11. Anastas PT, Warner JC (2000) Green chemistry: theory and practice. Oxford University Press, Oxford, pp 29–55
12. Ma SK, Gruber J, Davis C, Newman L, Gray D, Wang A, Grate J, Huisman GW, Sheldon RA (2010) A green-by-design biocatalytic process for atorvastatin intermediate. Green Chem 12:81–86. doi:10.1039/B919115C
13. Quarterly U.S. sales data for Lipitor. http://www.drugs.com/stats/lipitor. Accessed 1 May 2014
14. Winterton N (2001) Twelve more green chemistry principles. Green Chem 3:G73–G75. doi:10.1039/B110187K
15. Anastas PT, Zimmerman JB (2003) Design through the 12 principles of green engineering. Environ Sci Technol 37:94A–101A. doi:10.1021/es032373g
16. Tang SLY, Smith RL, Poliakoff M (2005) Principles of green chemistry: PRODUCTIVELY. Green Chem 7:761–762. doi:10.1039/B513020B
17. Tang S, Bourne R, Smith R, Poliakoff M (2008) The 24 principles of green engineering and green chemistry: "IMPROVEMENTS PRODUCTIVELY". Green Chem 10:268–269. doi:10.1039/B719469M
18. Anastas PT, Kirchoff MM (2002) Origins, current status, and future challenges of green chemistry. Acc Chem Res 35:686–694. doi:10.1021/ar010065m
19. Beach ES, Cui Z, Anastas PT (2009) Green chemistry: a design framework for sustainability. Energy Environ Sci 2:1038–1049. doi:10.1039/b904997p
20. Anastas P, Eghbali N (2010) Green chemistry: principles and practice. Chem Soc Rev 39:301–312. doi:10.1039/b918763b

21. Mulvihill MJ, Beach ES, Zimmerman JB, Anastas PT (2011) Green chemistry and green engineering: a framework for sustainable technology development. Annu Rev Environ 36:271–293. doi:10.1146/annurev-environ-032009-095500
22. Bourne RA, Poliakoff M (2011) Green chemistry: what is the way forward? Mendeleev Commun 21:235–238. doi:10.1016/j.mencom.2011.09.001
23. Horvath IT, Anastas PT (2007) Innovations and green chemistry. Chem Rev 107:2169–2173. doi:10.1021/cr078380v
24. Dichiarante V, Ravelli D, Albini A (2010) Green chemistry: state of the art through an analysis of the literature. Green Chem Lett Rev 3:105–113. doi:10.1080/17518250903583698
25. Sheldon RA (2008) E factors, green chemistry and catalysis: an odyssey. Chem Commun 29:3352–3365. doi:10.1039/b803584a
26. Constable DJC, Curzons AD, Cunningham VL (2002) Metrics to "green" chemistry—which are the best? Green Chem 4:521–527. doi:10.1039/b206169b
27. Constable DJC, Jimenez-Gonzalez C, Henderson RK (2007) Perspective on solvent use in the pharmaceutical industry. Org Process Res Dev 11:133–137. doi:10.1021/op060170h
28. Andraos J (2005) Unification of reaction metrics for green chemistry: applications to reaction analysis. Org Process Res Dev 9:149–163. doi:10.1021/op049803n
29. Information about the Presidential Green Chemistry Challenge. http://www2.epa.gov/green-chemistry/information-about-presidential-green-chemistry-challenge. Accessed 2 May 2014
30. Cann MC (1999) Bringing state-of-the-art, applied, novel, green chemistry to the classroom by employing the Presidential Green Chemistry Challenge Awards. J Chem Educ 76:1639–1641. doi:10.1021/ed076p1639
31. 2006 greener reaction conditions award. http://www2.epa.gov/green-chemistry/2006-greener-reaction-conditions-award. Accessed 1 May 2014
32. Elango V, Murphy MA, Smith BL, Davenport KG, Mott GN, Zey EG, Moss GL (1991) U.S. Patent 4981995; Lindley DD, Curtis TA, Ryan TR, de la Garza EM, Hilton CB, Kenesson TM (1991) U.S. Patent 5068448
33. Stuart NJ, Sanders AS (1968) U.S. Patent 3385886
34. Rainsford KD (2012) Ibuprofen: pharmacology, therapeutics and side effects. Springer, Basel
35. Drug record: ibuprofen. http://livertox.nlm.nih.gov/Ibuprofen.htm. Accessed 2 May 2014
36. Cann MC, Connelly ME (2000) Real world cases in green chemistry. ACS, Washington, DC 19–24
37. Green chemistry—the atom economy, student manual. Royal Society of Chemistry http://www.rsc.org/images/PDF1_tcm18-40521.pdf. Accessed 2 Feb 2014
38. Wang Z (2009) Comprehensive organic name reactions and reagents. Wiley, Hoboken, pp 841–845
39. Cann MC, Dickneider TA (2004) Infusing the chemistry curriculum with green chemistry using real-world examples, web modules, and atom economy in organic chemistry courses. J Chem Educ 81:977–980. doi:10.1021/ed081p977
40. Doble M, Kruthiventi AK (2007) Green chemistry and engineering. Academic Press, Elsevier Science and Technology Books
41. Desai AA (2011) Sitagliptin manufacture: a compelling tale of green chemistry, process intensification, and industrial asymmetric catalysis. Angew Chem Int Ed 50:1974–1976. doi:10.1002/anie.201007051
42. Kim D, Wang L, Beconi M, Eiermann GJ, Fisher MH, He H, Hickey GJ, Kowalchick JE, Leiting B, Lyons K, Marsilio F, McCann ME, Patel RA, Petrov A, Scapin G, Patel SB, Roy RS, Wu JK, Wyvratt MJ, Zhang BB, Zhu L, Thornberry NA, Weber AE (2005) (2R)-4-Oxo-4-[3-(Trifluoromethyl)-5,6-dihydro[1, 2, 4]triazolo[4,3-a]pyrazin-7(8H)-yl]-1-(2,4,5-trifluorophenyl) butan-2-amine: a potent, orally active dipeptidyl peptidase IV inhibitor for the treatment of type 2 diabetes. J Med Chem 48:141–151. doi:10.1021/jm0493156
43. Hansen KB, Balsells J, Dreher S, Hsiao Y, Kubryk M, Palucki M, Rivera N, Steinhuebel D, Armstrong JD III, Askin D, Grabowski EJJ (2005) First generation process for the preparation of the DPP-IV inhibitor sitagliptin. Org Process Res Dev 9:634–639. doi:10.1021/op0500786

44. Hansen KB, Hsiao Y, Xu F, Rivera N, Clausen A, Kubryk M, Krska S, Rosner T, Simmons B, Balsells J, Ikemoto N, Sun Y, Spindler F, Malan C, Grabowski EJJ, Armstrong JD III (2009) Highly efficient asymmetric synthesis of sitagliptin. J Am Chem Soc 131:8798–8804. doi:10.1021/ja902462q

45. Savile CK, Janey JM, Mundorff EC, Moore JC, Tam S, Jarvis WR, Colbeck JC, Krebber A, Fleitz FJ, Brands J, Devine PN, Huisman GW, Hughes GJ (2010) Biocatalytic asymmetric synthesis of chiral amines from ketones. Science 329:305–309. doi:10.1126/science.1188934

46. Balsells J, Hsiao Y, Hansen KB, Xu F, Ikemoto N, Clasuen A, Armstrong JD III (2010) Synthesis of sitagliptin, the active ingredient in Januvia® and Janumet®. In: Dunn PJ, Wells AS, Williams MT (eds) Green chemistry in the pharmaceutical industry. Wiley-VCH Verlag GmbH & Co, KGaA, Weinheim

47. Dunn PJ (2012) The importance of green chemistry in process research and development. Chem Soc Rev 41:1452–1461. doi:10.1039/c1cs15041c

48. 2006 greener synthetic pathways award. http://www2.epa.gov/green-chemistry/2006-greener-synthetic-pathways-award. Accessed 3 May 2014

49. 2010 greener reaction conditions award. http://www2.epa.gov/green-chemistry/2010-greener-reaction-conditions-award. Accessed 3 May 2014

50. Prat D, Pardigon O, Flemming H-W, Letestu S, Ducandas V, Isnard P, Guntrum E, Senac T, Ruisseau S, Cruciani P, Hosek P (2013) Sanofi's solvent selection guide: a step toward more sustainable processes. Org Process Res Dev 17:1517–1525. doi:10.1021/op4002565

51. Boethling R, Voutchkova A (2012) Handbook of green chemistry volume 9: designing safer chemicals, 1st edn. Wiley-VCH Verlag GmbH & Co. KGaA, Weinheim

52. Van Aken K, Strekowski L, Patiny L (2006) EcoScale, a semi-quantitative tool to select an organic preparation based on economical and ecological parameters. Beilstein J Org Chem 2. doi:10.1186/1860-5397-2-3

53. Taber GP, Pfisterer DM, Colberg JC (2004) A new and simplified process for preparing N-[4-(3,4-dichlorophenyl)-3,4-dihydro-1(2H)-naphthalenylidene]methanamine and a telescoped process for the synthesis of (1S-cis)-4-(3,4-dichlorophenol)-1,2,3,4-tetrahydro-N-methyl-1-naphthalenamine mandelate: key intermediates in the synthesis of sertraline hydrochloride. Org Process Res Dev 8:385–388. doi:10.1021/op0341465

54. Sarges R, Tretter JR, Tenen SS, Weissman A (1973) 5,8-disubstituted 1-aminotetralins. A class of compounds with a novel profile of central nervous system activity. J Med Chem 16:1003–1011. doi:10.1021/jm00267a010

55. Welch WM (1995) Discovery and preclinical development of the serotonin reuptake inhibitor sertraline. Adv Med Chem 3:113–148. doi:10.1016/S1067-5698(06)80005-2

56. Welch WM, Kraska AR, Sarges R, Koe BK (1984) Nontricyclic antidepressant agents derived from cis- and trans-1-amino-4-aryltetralins. J Med Chem 27:1508–1515. doi:10.1021/jm00377a021

57. Cann MC, Umile TP (2008) Real world cases in green chemistry, vol II. ACS, Washington, DC, pp 39–45

58. 2002 greener synthetic pathways award. http://www2.epa.gov/green-chemistry/2002-greener-synthetic-pathways-award. Accessed 3 May 2014

59. Alfonsi K, Colberg J, Dunn PJ, Fevig T, Jennings S, Johnson TA, Kleine HP, Knight C, Nagy MA, Perry DA, Stefaniak M (2008) Green chemistry tools to influence a medicinal chemistry and research chemistry based organisation. Green Chem 10:31–36. doi:10.1039/b711717e

60. Henderson RK, Jimenez-Gonzalez C, Constable DJC, Alston SR, Inglis GGA, Fisher G, Sherwood J, Binks SP, Curzons AD (2011) Expanding GSK's solvent selection guide—embedding sustainability into solvent selection starting at medicinal chemistry. Green Chem 13:854–862. doi:10.1039/c0gc00918k

Chapter 2
Atom Economy and Reaction Mass Efficiency

Abstract The green metrics atom economy (AE) and reaction mass efficiency (RME) are introduced and discussed. Following literature definitions, examples of reactions appropriate for upper-level undergraduate students are provided to illustrate how the metrics are calculated. In the case of atom economy, important assumptions regarding reactants, solvents and reagents are identified and explained. Several examples of inherently atom-efficient and inefficient reactions are also provided. In terms of reaction mass efficiency, the focus centers on a concise mathematical breakdown of various factors which contribute to changes in RME values in the context of two well-established definitions. A view of RME as a more robust metric that better captures the materials used during a chemical transformation is developed in the context of an undergraduate Suzuki reaction. With numerous academic and industrial examples comparing traditional syntheses with modern catalytic routes, the benefits and limitations of AE and RME are considered. Along with real-world case studies, the useful and effective application of these metrics is explained using several definitions of an ideal chemical reaction as points of reference. Finally, future projections and academic work are briefly outlined in order to highlight the development of these important metrics.

Keywords Atom economy · Reaction mass efficiency · Generalized reaction mass efficiency · Suzuki reaction · Product yield · Heterogeneous catalysis · Homogeneous catalysis · Biocatalysis

2.1 Atom Economy

2.1.1 Development and Motivation

The concept of atom economy (AE) was introduced in 1991 by Barry M. Trost at Stanford University [1]. In the past, the material efficiency of a chemical reaction was routinely quantified by measurement of the product yield, with an ideal value

© The Author(s) 2015
A.P. Dicks and A. Hent, *Green Chemistry Metrics*,
SpringerBriefs in Green Chemistry for Sustainability,
DOI 10.1007/978-3-319-10500-0_2

of 100 %. Atom economy has since sparked a "green" paradigm shift, as chemists began viewing reactions in terms of *how much* of the reactants are converted into a desired product. With the goal of achieving "synthetic efficiency in transforming readily available starting materials to the final target" [1], the primary motivation was to maximize the incorporation of reactant atoms into final products. This goal has led many chemists to focus their attention on adopting and developing processes that were inherently atom-efficient (Sect. 2.1.3). To help achieve higher selectivity and efficiency in organic syntheses, the application and development of catalytic systems was emphasized (Sect. 2.1.4). The development of theoretical aspects of atom economy has occurred both in isolation [2–5] and from the point of view of green metrics [6–8]. Collectively, these works illustrate the power of green metrics, their virtues and limitations, and their ability to promote innovation and change with regard to sustainable practice.

2.1.2 Definition and Key Assumptions

The ideal atom economy for a chemical transformation is taken as the process where all reactant atoms are found in the desired product [1]. In other words, atom economy is a calculation which measures "how much of the reactants remain in the final product" [6]. The percent atom economy of a generic stoichiometric chemical reaction to synthesize compound C is shown in Fig. 2.1. Since the calculation is essentially the molecular weight ratio of the final product divided by the sum of all reactants, it is possible to determine the atom economy for a reaction prior to undertaking any experimental work.

This calculation extends to a multi-step process where intermediates that are formed in one step and consumed during a later step are neglected (Fig. 2.2). Certain key assumptions about reactants, catalysts and reaction stoichiometry are necessary when calculating atom economy [6]. Firstly, a reactant is understood as any material that is incorporated into an intermediate or product during the synthesis. This includes protecting groups, catalysts used in stoichiometric quantities and acids or bases used for hydrolysis. Solvents, reagents or materials used in catalytic quantities are omitted from the analysis, as they do not contribute atoms to an intermediate and/or product.

$$A + B \xrightarrow[\substack{\text{Solvents} \\ \text{Catalysts}}]{\text{Reagents}} C + D$$

$$\% \text{ Atom Economy} = \frac{\text{GMW(C)}}{\text{GMW(A + B)}} \times 100\%$$

Fig. 2.1 Atom economy calculation for the synthesis of C

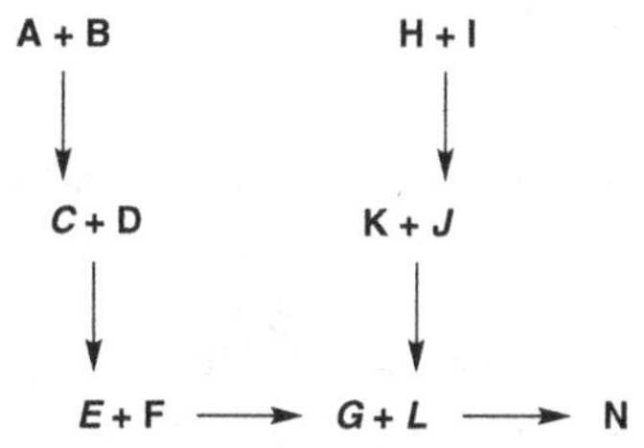

Fig. 2.2 Atom economy
calculations for products
G and N

$$A + B \qquad\qquad H + I$$
$$\downarrow \qquad\qquad\quad \downarrow$$
$$C + D \qquad\qquad K + J$$
$$\downarrow \qquad\qquad\quad \downarrow$$
$$E + F \longrightarrow G + L \longrightarrow N$$

$$\% \text{ Atom Economy of G} = \frac{\text{GMW(G)}}{\text{GMW(A + B + D + F)}} \times 100\%$$

$$\% \text{ Atom Economy of N} = \frac{\text{GMW(N)}}{\text{GMW(A + B + D + F + H + I + K)}} \times 100\%$$

A second assumption states that the chemical equation (which includes all starting materials and products) has been fully and correctly balanced. For example, if three equivalents of an inorganic base are consumed during a transformation such as the Suzuki reaction, three base equivalents must be included in the calculation of atom economy (Scheme 2.1) [9]. It is useful to think of atom economy in terms of accounting for consumed reactant material. Note that the calculation does not reflect actual experimental masses and volumes. Moreover, knowledge of reaction mechanisms is highly recommended. Many of these elements are discussed in greater detail elsewhere [5–7].

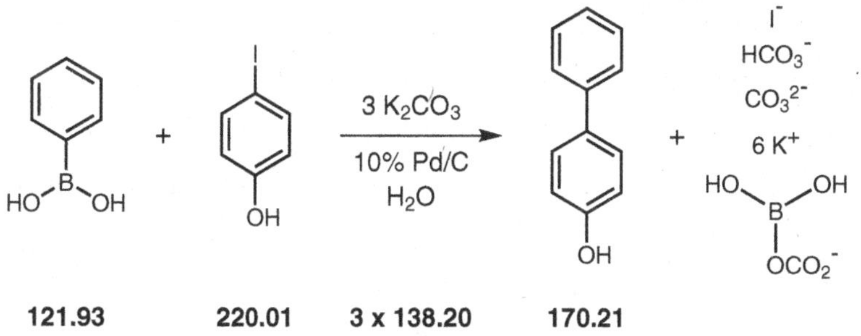

121.93   220.01   3 x 138.20   170.21

$$\% \text{ Atom Economy} = \frac{170.21}{121.93 + 220.01 + 3 \times 138.20} \times 100\% = 22.5\%$$

Scheme 2.1 Atom economy of a balanced Suzuki reaction

2.1.3 *Reaction Types: The Good, the Bad and the Ugly*

Atom economy calculations generally show that addition and rearrangement reactions are preferred over substitutions and eliminations. The mercury (II)-catalyzed hydration of alkynes and the benzilic acid rearrangement are examples of 100 % atom-efficient reactions (Scheme 2.2) [10].

Although rearrangements often proceed with perfect atom economy, certain addition reactions do not. Examples include an osmium tetroxide-mediated dihydroxylation and a Simmons-Smith cyclopropanation reaction (Scheme 2.3) [11]. In particular, the mechanism of the Simmons-Smith reaction shows that a significant portion of the starting materials ends up as molecular waste [12]. This inefficiency provides opportunities for designing new reactions with the goal of improving atom economy.

For example, the Upjohn dihydroxylation uses *N*-methylmorpholine *N*-oxide (NMO) as a cheap co-oxidant to render the toxic and expensive osmium tetroxide catalytic via re-oxidation (Scheme 2.4) [11, 13]. In a recent article, Maurya et al. described a new catalyst-free cyclopropanation which uses electron-deficient alkenes (Scheme 2.5) [14]. In this reaction, the increased electrophilicity of doubly-activated alkenes facilitates a Michael-induced ring closure with ethyl diazoacetate. By eliminating stoichiometric reagents and minimizing waste, a much higher atom economy is achieved. Addition reactions are therefore excellent candidates for designing more atom-efficient processes [3–5, 8, 15].

Although substitution and elimination reactions are intrinsically wasteful (Scheme 2.6), there exists opportunities to design for better atom economy. For

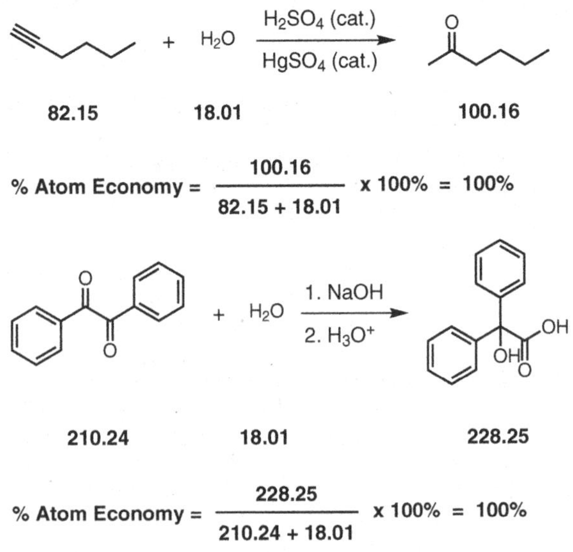
+ H2O
H2SO4 (cat.)
HgSO4 (cat.)
O
82.15
18.01
100.16
% Atom Economy = 100.16 / 82.15 + 18.01 x 100% = 100%
O
O
+ H2O
1. NaOH
2. H3O+
OH
OH
O
210.24
18.01
228.25
% Atom Economy = 228.25 / 210.24 + 18.01 x 100% = 100%

Scheme 2.2 Atom economy calculations for the Hg^{2+}-catalyzed hydration of an alkyne and a benzilic acid rearrangement

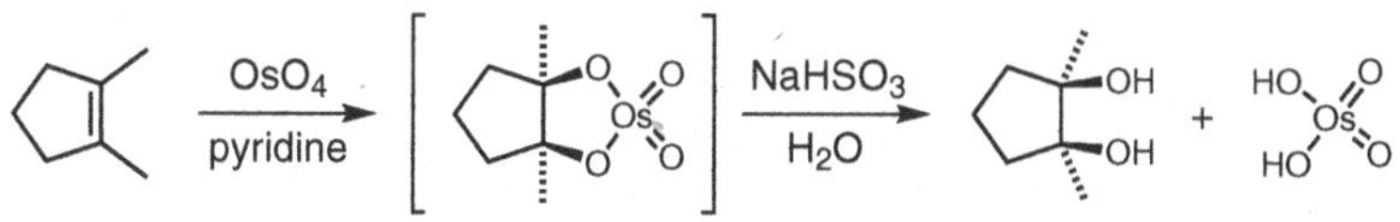

97.17   254.23   2 x 18.01   130.19

$$\% \text{ Atom Economy} = \frac{130.19}{97.17 + 254.23 + 2 \times 18.01} \times 100\% = 34\%$$

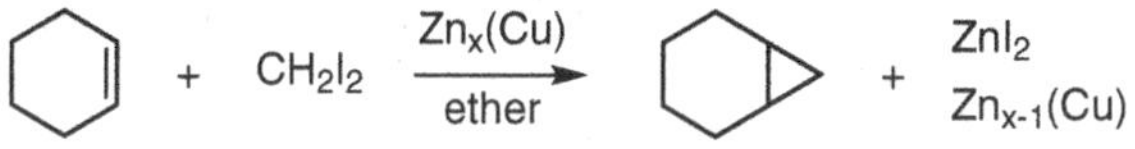

82.15   267.84   128.93   96.17

$$\% \text{ Atom Economy} = \frac{96.17}{82.15 + 267.84 + 128.93} \times 100\% = 20\%$$

Scheme 2.3 Atom economy calculations for an osmium tetroxide-mediated dihydroxylation and a Simmons-Smith cyclopropanation

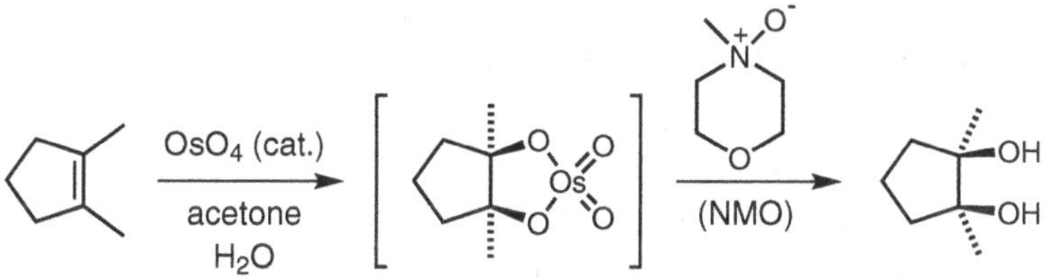

97.17   2 x 18.01   117.15   130.19

$$\% \text{ Atom Economy} = \frac{130.19}{97.17 + 2 \times 18.01 + 117.15} \times 100\% = 52\%$$

Scheme 2.4 Improved atom economy of an OsO$_4$-mediated dihydroxylation using *N*-methyl-morpholine as a co-oxidant

example, the preparation of alkyl halides from alcohols is routinely effected with either phosphorus tribromide (PBr$_3$, Scheme 2.6) or thionyl chloride (SOCl$_2$, Scheme 2.7) [16]. Choosing the appropriate substitution involves deciding between gaining access to a better leaving group (Br) for further reaction, or including a step with a higher atom economy. Recent work in substitution reactions has shown that catalytic conditions can also improve atom economy [17, 18].

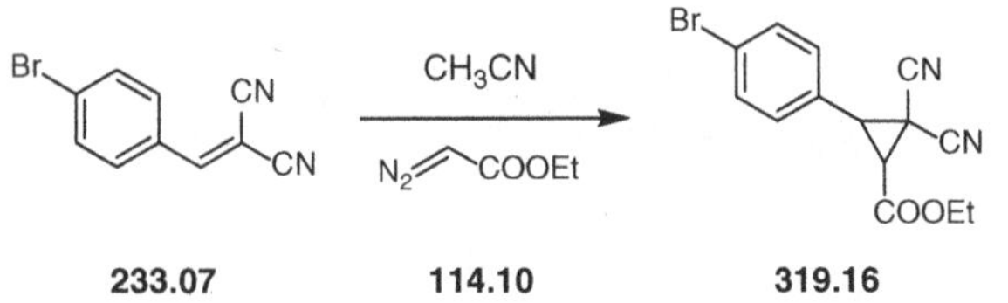

$$\% \text{ Atom Economy} = \frac{319.16}{233.07 + 114.10} \times 100\% = 92\%$$

Scheme 2.5 Atom-economic catalyst-free cyclopropanation of an electron deficient alkene with ethyl diazoacetate

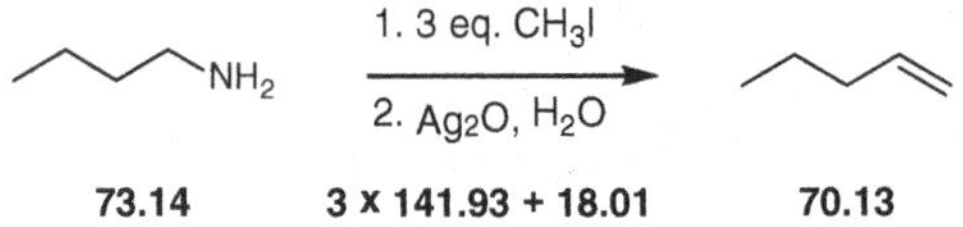

$$\% \text{ Atom Economy} = \frac{70.13}{73.14 + 3 \times 141.93 + 18.01} \times 100\% = 14\%$$

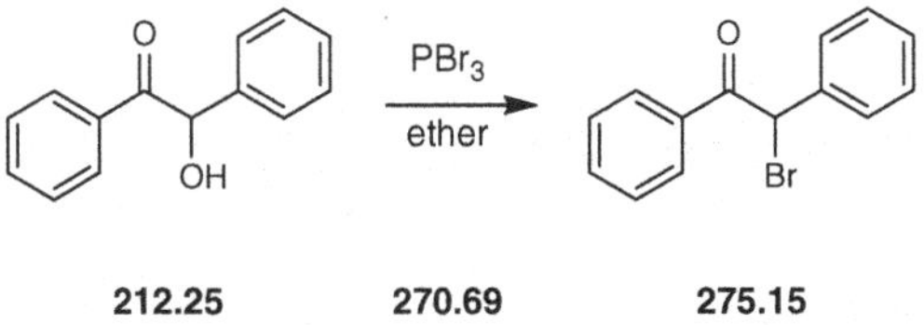

$$\% \text{ Atom Economy} = \frac{275.15}{212.25 + 270.69} \times 100\% = 57\%$$

Scheme 2.6 Atom economy calculations for a Hofmann elimination and bromination of a secondary alcohol with PBr_3

Scheme 2.7 Atom economy calculation for chlorination of a secondary alcohol with $SOCl_2$

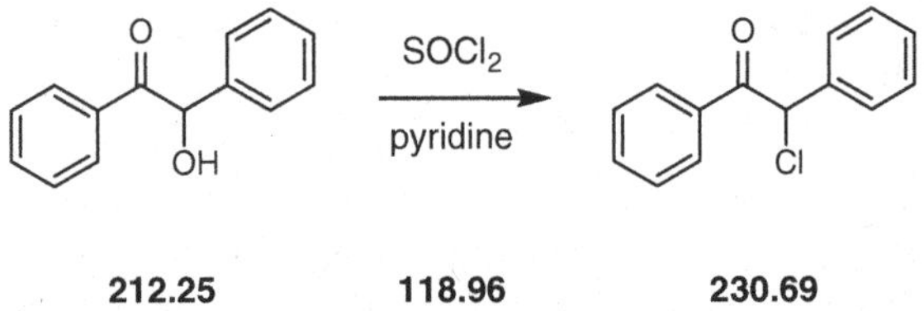

$$\% \text{ Atom Economy} = \frac{230.69}{212.25 + 118.96} \times 100\% = 70\%$$

2.1.4 Catalysis, Industry and Innovation

Reaction catalysis is generally understood in terms of kinetics, with an emphasis on the enhanced rate of a chemical process in the presence of a regenerated catalyst. Because they are not consumed, catalysts are omitted from the formal calculation of atom economy. Catalysts work by providing an alternative reaction pathway involving lower energy transition states and a lower activation barrier for the reaction rate-determining step [19]. Catalysis can therefore promote atom-efficient reactions which are otherwise energetically disfavoured. With the use of hetero-geneous, homogeneous and biocatalytic strategies, it is possible to reduce (if not eliminate) experimental constraints such as extra synthetic steps, stoichiometric components and energy inputs. Consequently, it is possible to undertake not just greener syntheses, but ones with higher atom economies.

2.1.4.1 Heterogeneous Catalysis

A process where a catalyst in one phase (usually a solid) interacts with reactants in a different phase (usually a gas or liquid) is called *heterogeneous catalysis*. This interaction occurs via adsorption of reactants onto the surface of the catalyst. A fuller discussion of the principles of heterogeneous catalysis has been published elsewhere [19, 20]. In many sources, examples of applied heterogeneous catalysis on an industrial scale frequently cite the preparation of ethylene oxide [21, 22] and the nickel-catalyzed hydrogenation of nitrobenzene [22, 23] (Scheme 2.8). In the case of nitrobenzene, the original process had an atom economy of 35 %. When combined with the loss of valuable iron-containing reactants, it was clear that a cheaper more efficient process was necessary to accommodate the global demand for aniline. To help solve this problem, nickel was picked as a cheap, robust and easily recyclable heterogeneous catalyst for the production of aniline. The new nickel-catalyzed process thus achieved an atom economy of 72 %. The synthesis of aniline has also proven valuable from a pedagogical perspective. In a recent article, Mercer et al. have described a student-driven multi-metric analysis of five different routes towards the production of aniline from benzene [24].

In 2007, the American Chemical Society Green Chemistry Institute Pharma-ceutical Roundtable (ACS GCI PR) created a research agenda to promote areas of mutual interest for advancing green chemistry principles [25]. One area requiring significant improvement was amide bond formation, with a special emphasis on the need to eliminate inefficient reagents such as carbodiimides, and phosphonium/uranium salts, among many others. The inefficiency of many amide bond coupling reagents, their tendency to form toxic or corrosive by-products, and their costly waste streams have been discussed in the literature [26]. Since the Roundtable's findings, numerous solutions to this problem have been proposed [27]. A promising approach emerged in 2009 [28] featuring the use of thermally-activated K60 silica as an affordable, readily available and benign heterogeneous catalyst (Scheme 2.9).

Traditional Béchamp Process:

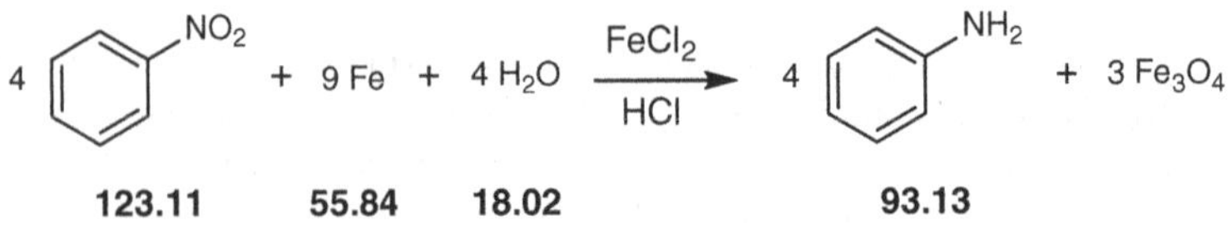

$$\% \text{ Atom Economy} = \frac{4 \times 93.13}{4 \times 123.11 + 9 \times 55.84 + 4 \times 18.02} \times 100\% = 35\%$$

Nickel Catalyzed Hydrogenation Process:

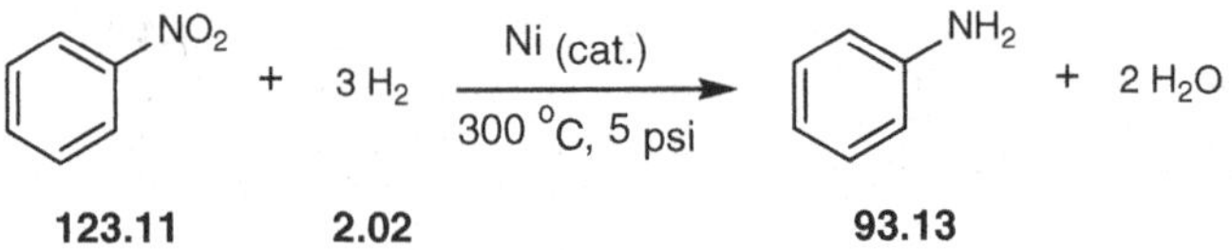

$$\% \text{ Atom Economy} = \frac{93.13}{123.11 + 3 \times 2.02} \times 100\% = 72\%$$

Scheme 2.8 Atom economy calculations for the traditional Béchamp process and the nickel-catalyzed hydrogenation of nitrobenzene

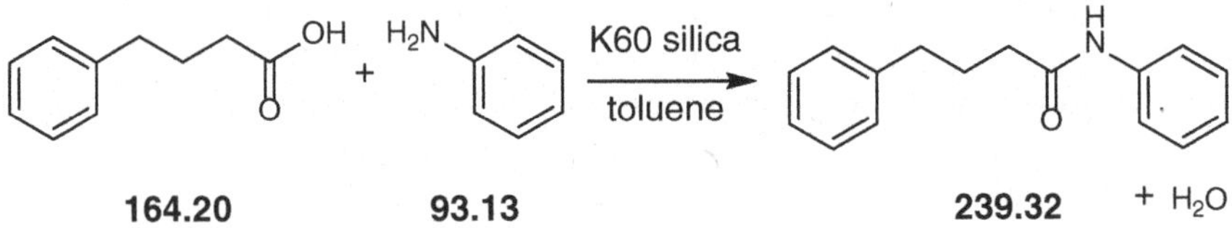

$$\% \text{ Atom Economy} = \frac{239.32}{164.20 + 93.13} \times 100\% = 93\%$$

Scheme 2.9 Atom economy calculation for the K60 silica catalyzed synthesis of 4, N-diphenylacetamide

Although high temperatures were required to prevent the product from becoming trapped within silica pores, the reaction between 4-phenylbutanoic acid and aniline produced a yield of 74 % with an atom economy of 93 %. Furthermore, the authors carried out continuous flow experiments to demonstrate both catalyst recyclability and reaction completion on scales required for industrial applications. On a different note, an important subclass of heterogeneous catalysis is phase-transfer catalysis. Phase-transfer catalysts work to adsorb reactants and transfer them between liquid phases to promote reactivity. Recent work outlined use of quaternary ammonium salts as benign reusable pseudo-phase-transfer catalysts for a benzoin condensation carried out in water, which occurs with 100 % atom economy (Scheme 2.10) [29].

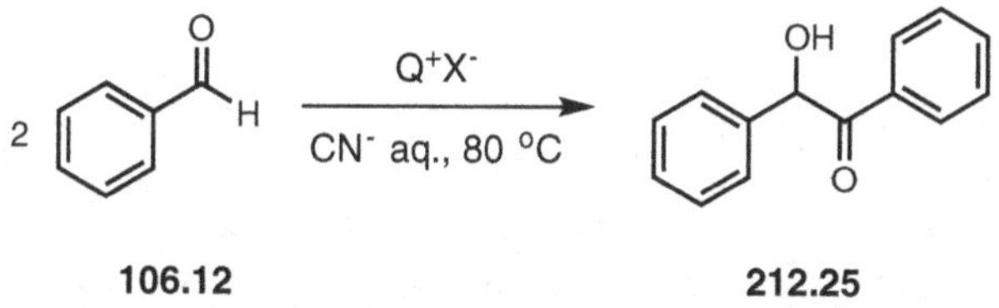

$$\% \text{ Atom Economy} = \frac{212.25}{2 \times 106.12} \times 100\% = 100\%$$

Scheme 2.10 Atom economy calculation for a benzoin condensation catalyzed by a quaternary ammonium salt (Q^+X^-)

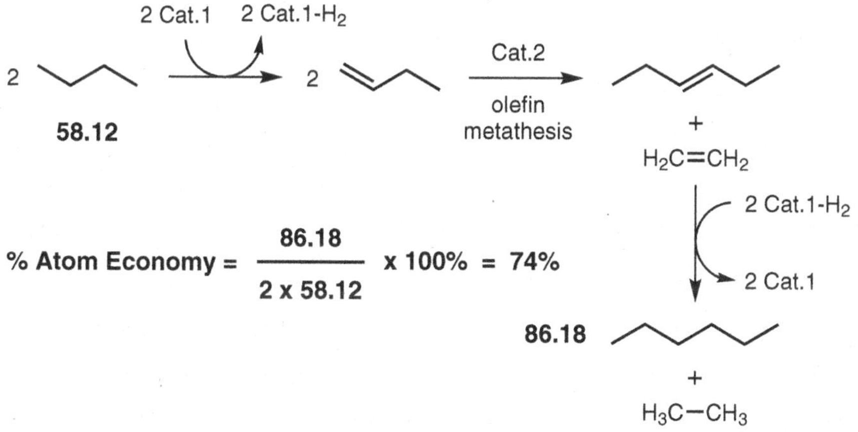

$$\% \text{ Atom Economy} = \frac{86.18}{2 \times 58.12} \times 100\% = 74\%$$

Scheme 2.11 Atom economy calculation for the metathesis of butane using a dual-catalyst system

Another approach in applying heterogeneous catalysis involves combining multiple catalysts in a single system. In 2006, Goldman et al. used two catalysts to carry out the metathesis of *n*-alkanes (Scheme 2.11) [30]. A "pincer"-ligated iridium complex was used as the hydrogen transfer catalyst to effect both alkane dehydrogenation and olefin hydrogenation. A standard solid phase catalyst was used for olefin metathesis. The high atom economy and selectivity achieved by this system makes the approach very elegant. Current research in heterogeneous catalysis is aimed at designing recyclable high-selectivity catalysts for reactions requiring C–H activation [31].

2.1.4.2 Homogeneous Catalysis

Homogeneous catalysis takes place in a system where reactants and catalysts are found in the same phase (usually both liquids). As distinguished from a heterogeneous process which involves surface chemistry, homogeneous catalysis proceeds via discrete association and dissociation steps within solution. For simplicity one can separate homogeneous catalysis into two categories: those involving metal complexes, and others without metals. The former is often referred to as organometallic catalysis while the latter involves acid/base catalysis and organocatalysis. For more details on this topic the reader is referred to an excellent book written by Rothenberg [19].

Although heterogeneous catalysis is applied in nearly 90 % of industrial processes, homogeneous catalysis is gaining momentum [19]. The 1970 du Pont adiponitrile synthesis catalyzed by a nickel-tetrakis(phosphite) complex is an example of a major industrial process occurring with 100 % atom economy (Scheme 2.12) [32, 33]. Moreover, in terms of advancing green chemistry principles, an article by Allen and Crabtree effectively demonstrates ways to improve upon catalytic systems that are already deemed green [34]. Along the same lines, the traditional approach to β-alkylation of alcohols consists of a three-step atom inefficient route (Scheme 2.13). Earlier work by Crabtree's group had identified

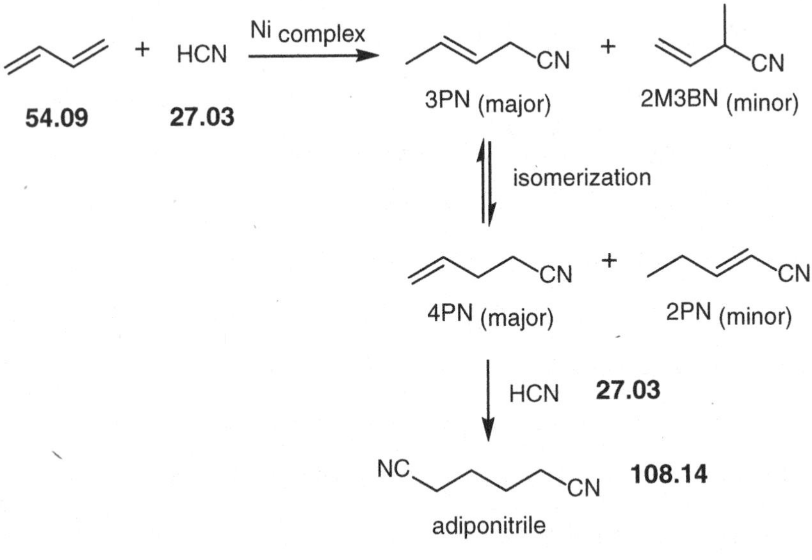

$$\% \text{ Atom Economy} = \frac{108.14}{54.09 + 2 \times 27.03} \times 100\% = 100\%$$

Scheme 2.12 Atom economy calculation for the nickel-catalyzed synthesis of adiponitrile

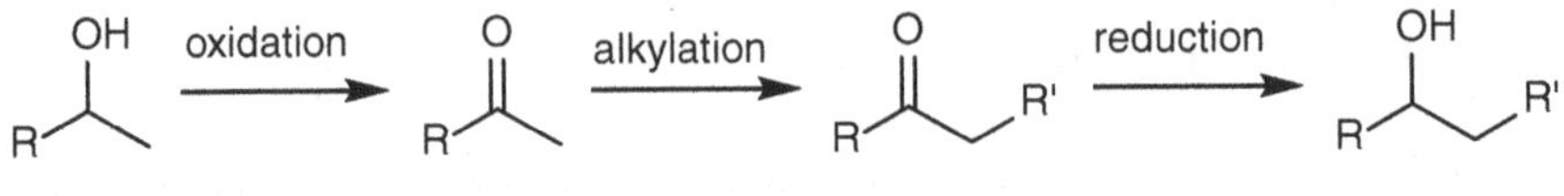

Scheme 2.13 Traditional three-step β-alkylation of alcohols

homogeneous iridium and ruthenium catalyst complexes giving a one-pot β-alkylation with an atom economy of 62 % [35, 36].

   In a recent article, Crabtree et al. used an alkali metal base to catalyze the same β-alkylation (Scheme 2.14) [34]. The proposed mechanism starts with an Oppenauer oxidation in air, followed by a base-catalyzed aldol reaction, and ending with a Meerwein–Ponndorf–Verley-type reduction [37]. Aside from an improved atom economy, the method demonstrated a lower energy input and use of cheaper, less toxic earth metals as opposed to transition metals.

   Finally, an important technique gaining traction in homogeneous catalysis is the idea of "hydrogen borrowing". This approach uses catalysts as carriers of hydrogen atoms to promote redox-neutral reactions such as an alcohol-amine coupling (Scheme 2.15). Following this strategy, researchers made a GlyT1 inhibitor for the treatment of schizophrenia on a kilogram scale using an iridium complex as the catalyst [38]. Although many catalytic systems break down when involved in scale-up,

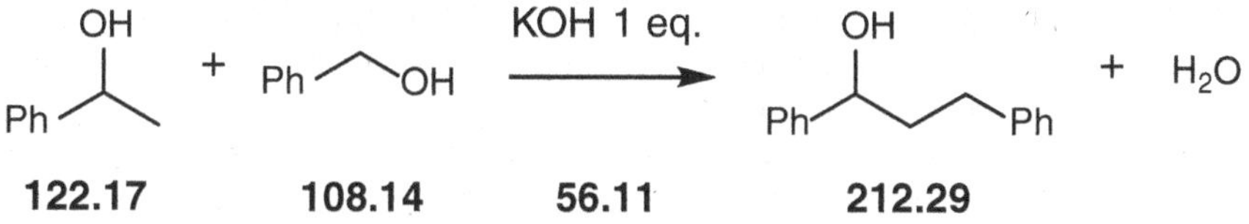

$$\% \text{ Atom Economy} = \frac{212.29}{122.17 + 108.14 + 56.11} \times 100\% = 74\%$$

Scheme 2.14 Atom economy calculation for base-catalyzed β-alkylation of a secondary alcohol

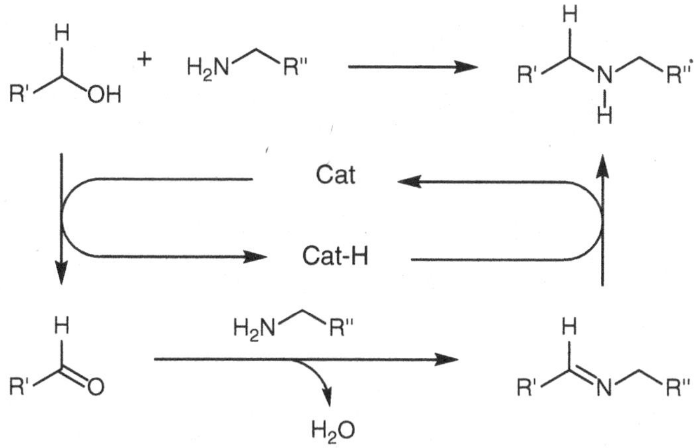

Scheme 2.15 Redox-neutral alcohol-amine coupling using hydrogen borrowing. Adapted with permission from [38]. Copyright 2011 American Chemical Society

this atom-efficient process working at near industrial scale demonstrates the versatility afforded by homogeneous catalysis.

2.1.4.3 Biocatalysis

Biocatalysis requires using enzymes to promote chemical reactions. Although enzymes have numerous green chemistry advantages including biodegradability, safety and high selectivity, it has been estimated that only about 130 routes have been commercialized as of 2002 [39]. This figure has steadily risen due to advances in recombinant DNA technology, protein engineering and immobilization methods that make the production, manipulation and optimization of enzymes economically feasible [40, 41].

In terms of atom economy, the manufacture of 6-aminopenicillanic acid (6-APA) from penicillin G highlights the power of biocatalysis. 6-APA is an important precursor to penicillin and cephalosporin antibiotics and has been traditionally made by a four-step deacylation process (Scheme 2.16) [42].

The route involved silyl protection of the penicillin G carboxyl group, transformation of the secondary amide moiety into an imine chloride with phosphorous pentachloride, enol ether formation, and finally hydrolysis leading to an overall atom economy of 28 %. In a 2001 review, Sheldon et al. explained the development of the biocatalytic process which uses a stable penicillin G acylase enzyme having an atom economy of 58 % (Scheme 2.17) [43]. Owing to a dramatic reduction in

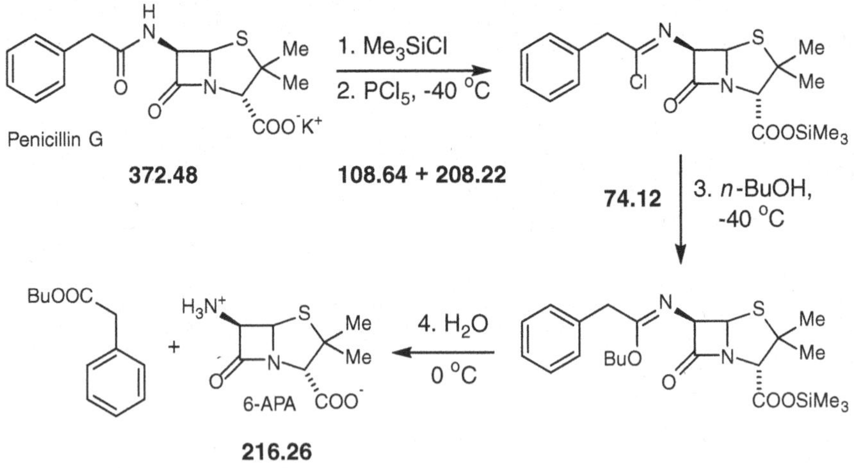

$$\% \text{ Atom Economy} = \frac{216.26}{372.48 + 108.64 + 208.22 + 74.12} \times 100\% = 28\%$$

Scheme 2.16 Atom economy calculation for the traditional 4-step deacylation of penicillin G to 6-aminopenicillanic acid (6-APA)

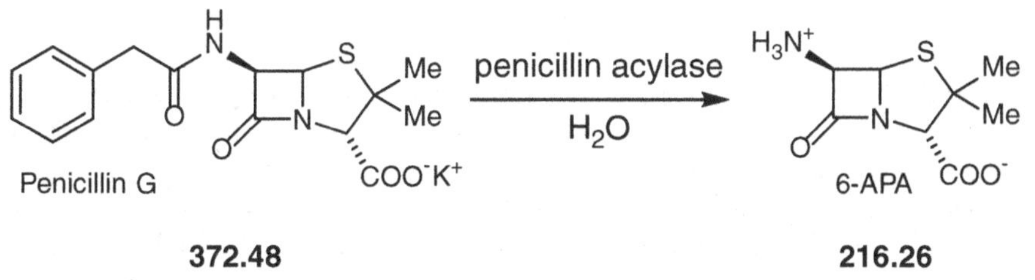

Scheme 2.17 Atom economy calculation for the penicillin acylase-catalyzed production of 6-aminopenicillanic acid (6-APA)

waste as well as milder reaction conditions, it was explained that the biocatalytic process had completely replaced the traditional deacylation route. Similar examples available in other publications emphasize the greenness of biocatalysis [44, 45], as well as its increased use in industry [46, 47].

2.1.5 100 % Atom Economy: Above and Beyond

Since atom economy reflects the intrinsic efficiency of a balanced chemical reaction, it is often conceptually isolated from the wider goals of synthesis and green chemistry. It should be stressed that an ideal atom economy should not deter one from considering other important reaction components, including yield, solvent use, catalyst recovery, energy and toxicity. Rather, an ideal atom economy should be the ultimate goal and the selection standard for achieving the greenest possible process.

As an example, the DuPont adiponitrile synthesis has an atom economy of 100 % (Scheme 2.12). However, the process depends on an equilibrium which favours the formation of the more thermodynamically stable 2-pentenenitrile (2PN). The equilibrium ratio of 3PN/2PN/4PN is 20:78:1.6 respectively [33]. Fortunately, 4PN is the favourable kinetic product [48], and the catalyst ligands can be made bulky to promote formation of the linear 3PN in order to ultimately form adiponitrile in 98 % yield [49]. With a longer reaction time the product yield would be lower.

On many occasions the reaction yield determines the most efficient catalyst and the optimal reaction conditions to be used. A recent article investigating the microwave-assisted multicomponent synthesis of quinolines illustrated the importance of reaction yield [50]. The synthesis of 2,4-diphenylquinoline was shown to be closely dependent on the acidic nature of the catalyst, the temperature inside the microwave, and the reaction time, giving product yields between 10–96 %. Since the atom economy remains fixed regardless of the chosen catalyst and other reaction parameters, it is important to recognize that an atom economy analysis will not always determine the greenest approach.

Although the chemical industry places great emphasis on reaction yield, a multi-metric analysis is often more appropriate for studying the efficiency and greenness of a synthesis. The 2009 article describing K60 silica-catalyzed amide bond forming reactions illustrates this point [28]. Included is a multi-metric comparison between four catalysts in the synthesis of 4,N-diphenylbutyramide. Accordingly, this approach shows that due to a high atom economy and a tenfold reduction in overall waste (as measured by the E factor), the efficiency of the K60 silica catalyzed process overshadows its good product yield.

In concluding the section on atom economy, it is appropriate to consider Trost's 1995 statement concerning his hopes for the future of green chemistry and the metric he proposed. "As the legitimate concerns of society for wise use of our limited resources with minimal environmental risk grow, the ability to produce the chemicals needed to improve the human condition will hinge on the inventfulness of chemists to design more efficient syntheses" [2].

2.2 Reaction Mass Efficiency (RME)

2.2.1 History and Development

The first article of the journal *Green Chemistry* (published in 1999) outlined the importance of metrics in identifying and meeting the challenges of sustainability [51]. This paper was significant as it marked the first time the term "mass efficiency" was used to describe green practices. One year later, Steinbach and Winkenbach introduced the term "balance yield" (synonymous with mass efficiency) as a measure of productivity [52]. Calculated as "main product amount" divided by "balance sheet total input", the balance yield was deemed important as it emphasized productivity, rather than waste, as "a key technical goal in industrial production" [52]. This new perspective marked the birth of globally-oriented mass based metrics which accounted for both the intrinsic and experimental aspects of a chemical reaction. Shortly after these developments, chemists introduced metrics such as mass index [53] (known today as process mass intensity, Chap. 3) as well as clearly-defined equations for reaction mass efficiency [6, 8, 54–61].

2.2.1.1 A Good Start: The Curzons Definition

In 2001, researchers from GlaxoSmithKline (GSK) presented a list of green metrics used by their company to promote sustainable development [54]. Among these, reaction mass efficiency (RME) was emphasized as a realistic metric for describing the greenness of a process. Calculated as product mass divided by the sum of the masses of reactants appearing in the balanced chemical equation [6, 54], it was eventually recognized that RME accounts for yield, stoichiometry and atom economy. This important connection can be drawn by considering a generic reaction

$$A \quad + \quad B \quad \xrightarrow{\substack{\text{reagents} \\ \text{solvents} \quad \cdot \\ \text{catalysts}}} \quad C$$

Mass:	m_1	m_2	m_3
Moles:	x	y	z
GMW:	MW_1	MW_2	MW_3

Fig. 2.3 Generic addition reaction where it is assumed that reactant B is in excess (i.e. y > x)

$$
\begin{aligned}
\text{RME} &= \frac{m_3}{m_1 + m_2} = \frac{z(MW_3)}{x(MW_1) + y(MW_2)} \\[2mm]
&= \frac{z(MW_3)}{x(MW_1) + (y - x + x)(MW_2)} = \frac{z(MW_3)}{x(MW_1) + x(MW_2) + (y - x)(MW_2)} \\[2mm]
&= \frac{z}{x} \cdot \frac{MW_3}{MW_1 + MW_2 + x^{-1}(y - x)MW_2} \cdot \frac{(MW_1 + MW_2)^{-1}}{(MW_1 + MW_2)^{-1}} \\[2mm]
&= \frac{z}{x} \cdot \frac{MW_3}{MW_1 + MW_2} \cdot \frac{1}{1 + \dfrac{(y - x)MW_2}{x(MW_1 + MW_2)}} \\[2mm]
&= \text{Yield} \; \cdot \; \text{Atom Economy} \; \cdot \; \frac{1}{\text{Stoichiometric Factor}}
\end{aligned}
$$

Fig. 2.4 Derivation of reaction mass efficiency as the product of yield, atom economy and the inverse of a stoichiometric factor according to the Curzons definition

(Fig. 2.3) where an excess amount of reactant B is added to reactant A to form product C. Using this example, the Curzons reaction mass efficiency is derived in Fig. 2.4.

The Curzons RME is therefore equivalent to a mathematical product consisting of yield, atom economy and the inverse of a stoichiometric factor, a term introduced by Andraos to measure reactant excess [57–59]. It is worthwhile to check this derivation via Curzons' own example (Scheme 2.18) [54].

Here, it is given that 10.81 g of benzyl alcohol reacts with 21.9 g of p-toluenesulfonyl chloride to form 23.6 g of the sulfonate ester. Applying the equation derived in Fig. 2.4 results in a matching RME value of 0.72 for the reaction. Note that the values of yield, atom economy, stoichiometric factor and RME are expressed in absolute form (i.e. as a value ranging between 0 and 1). This is done to make the RME product meaningful. Percent values cannot achieve this, and are therefore omitted for the remainder of the chapter. Understanding the Curzons RME as a product of distinct terms is important as it allows for developing more rational optimization strategies when deciding how to improve the greenness of a process.

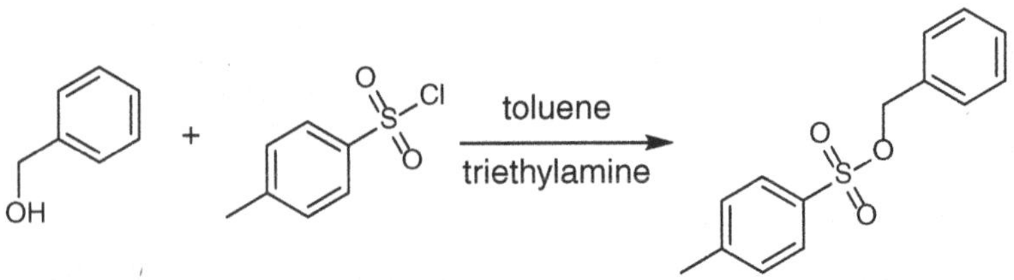

Mass: 10.81 g   21.9 g   23.6 g
Moles: 0.10   0.115   0.09
GMW: 108.14   190.64   262.32

$$RME = \frac{23.6}{10.81 + 21.9} = 0.72 \qquad\qquad Yield = \frac{0.09}{0.10} = 0.90$$

$$Atom\ Economy = \frac{262.32}{190.64 + 108.14} = 0.88$$

$$Stoichiometric\ Factor = 1 + \frac{(0.115 - 0.10) \times 190.64}{0.10 \times (108.14 + 190.64)} = 1.10$$

$$RME = Yield \cdot AE \cdot \frac{1}{SF} = 0.90 \times 0.88 \times (1.10)^{-1} = 0.72$$

Scheme 2.18 Two calculations of the Curzons RME for the esterification of benzyl alcohol and *p*-toluenesulfonyl chloride [52]

Curzons et al. further justified their metric using cost comparison models for drug manufactures at GSK [6]. Using this data they showed that atom economy influences the manufacture cost of pharmaceuticals much less than yield and stoichiometry. Developing a simple method to account for all three variables thus proved valuable for GSK. In subsequent years, Curzon's RME has gained greater appreciation from chemists in research, commercial and educational settings [55–61].

2.2.1.2 A Unifying Concept: The Andraos Definition

A crucial development of the reaction mass efficiency concept came in 2005 through the work of Andraos [57, 58]. In his mathematical treatise on green metrics [57–59], Andraos recognized that mass efficiency should account for all the materials involved in a chemical process, and not simply the reactant and product masses. This includes the mass of catalysts, solvents, and work-up/purification materials. As a result, Andraos proposed a generalized reaction mass efficiency formula which is simply the mass of the desired product divided by the total mass of all input material relevant to the reaction [57]. Using a similar derivation to that shown in Fig. 2.4, one can see that the generalized RME for a reaction can be broken down into a product of yield, atom economy, the stoichiometry factor and a new component called the material recovery parameter (introduced to account for catalysts, solvents, and work-up/purification materials [59]). Isolating these four terms in absolute form makes it possible to optimize for efficiency by seeking to achieve values of unity for each parameter.

In addition, by placing assumptions on the recovery and recyclability of reaction components, it became possible to identify context-specific RME formulae for a particular process [59]. In this context, the best-case scenario assumes complete recovery of excess reagents, catalysts, solvents and work-up materials. Consequently, the RME (known as the maximum or kernel RME) is simply the product of the reaction yield and atom economy [57, 59]. At the next level, stoichiometry is considered and as a result one obtains the Curzons RME (Fig. 2.4). The remaining RME formulae [59] follow from the various permutations of recovering one or more of the reaction components mentioned previously. These subtleties are highlighted using the Suzuki reaction mentioned at the start of the chapter (Scheme 2.19).

The experimental conditions shown follow a literature procedure and include an average student yield of 67 %, as obtained in a third year undergraduate lab course at the University of Toronto [9, 62]. In this example, the kernel RME is identical to the Curzons RME which means that reagent excess is not present. When the masses of remaining reaction components are included, the RME decreases significantly to 0.0023. To provide some perspective, Fig. 2.5 shows the mass percent distribution

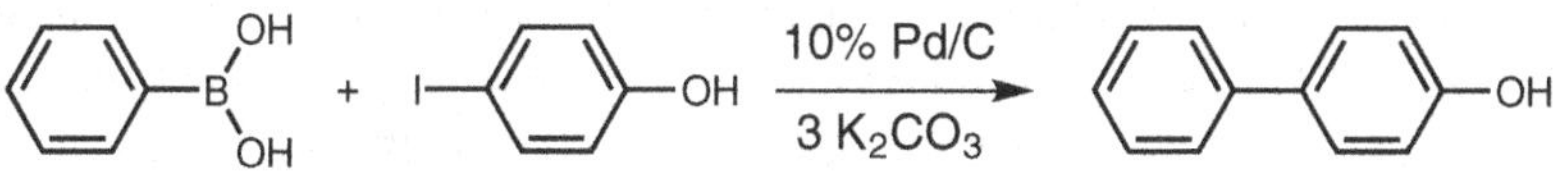

Mass:	0.122 g	0.220 g	0.415 g	0.115 g
Moles:	1.00 mmol	1.00 mmol	3.00 mmol	0.675 mmol
GMW:	121.93	220.01	414.6	170.21

Catalyst Mass: 0.003 g   Reaction Solvent Mass: 11 g

Work-up/Purification Material Mass: 38.1 g

Atom Economy = 0.225   Yield = 0.675

Kernel RME = 0.225 x 0.675 = 0.152

$$\textbf{Curzons RME} = \frac{0.115}{0.122 + 0.220 + 0.415} = 0.152$$

$$\textbf{Generalized RME} = \frac{0.115}{0.122 + 0.220 + 0.415 + 0.003 + 11 + 38.1} = 0.0023$$

Scheme 2.19 Reaction mass efficiency calculations for the Suzuki reaction using typical results obtained by undergraduate third year students at the University of Toronto [9, 62]

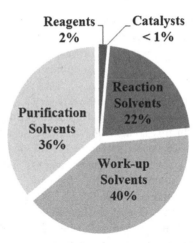

Fig. 2.5 Mass distribution profile of a Suzuki reaction

profile of the Suzuki reaction. With solvents occupying 98 % of the mass involved in the experiment, the decrease between the Curzons RME and the generalized RME becomes understandable.

Although previous reports on the Suzuki reaction demonstrated the recycling of the palladium catalyst [63], catalyst recovery has a negligible effect on the generalized RME. On the other hand, eliminating all solvents would increase the RME by a factor of 65. This analysis highlights the opportunities for optimizing the efficiency of the Suzuki reaction. Finally, the Andraos equations become valuable when experimental masses are not reported in the literature. Expressing the RME as a mathematical product involving reaction yield (routinely reported) and atom economy (easily calculable) enables a simple evaluation when determining the merits for why or why not a certain process might be considered green.

2.2.2 Applying RME to Catalysis

The synthesis of 2,4-diphenylquinoline (Sect. 2.1.5) showed that atom economy is limited in its potential to measure process efficiency [50]. Improving an atom economy of 93 % given reaction yields below 50 % will accomplish little in terms of achieving productivity and efficiency. In this section, the reaction mass efficiency metric will be used to provide a more global and robust perspective on greenness and sustainability.

2.2.2.1 Mass Efficiency in Heterogeneous Catalysis

In a recent article, Mercer et al. analyzed five industrial routes for the conversion of benzene into aniline [24]. The fifth and most efficient process (Scheme 2.20) included two steps: nitration of benzene via electrophilic substitution followed by

Mass: 589.7 kg   480.8 kg   907.2 kg (98% yield)
GMW: 78.11   63.01   123.11

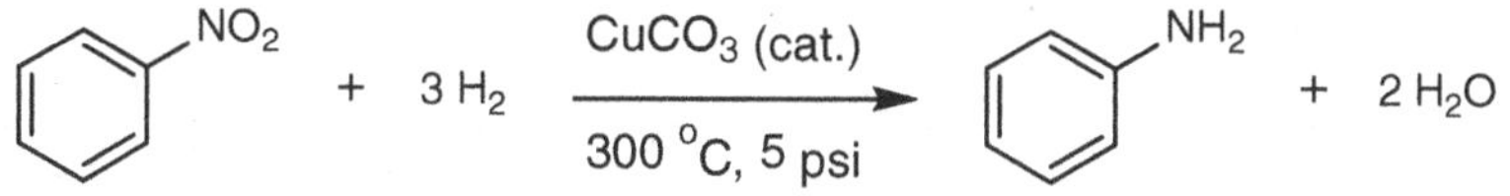

Mass: 907.2 kg   >43.6 kg   671.3 kg (98% yield)
GMW: 123.11   6.05   93.13

Catalyst Mass: 5.03 kg   Reaction Solvent Mass: 654.1 + 98.9 = 753 kg
Work-up and Purification Material Mass: 9.1 kg

Atom Economy = 0.63   Yield = 0.96

Kernel RME = 0.63 x 0.96 = 0.605

$$\text{Curzons RME} = \frac{671.3}{589.7 + 480.8 + 43.6} = 0.603$$

$$\text{Generalized RME} = \frac{671.3}{589.7 + 480.8 + 43.6 + 5.03 + 753 + 9.1} = 0.357$$

Scheme 2.20 Reaction mass efficiency calculations for the industrial conversion of benzene to aniline [24]

the familiar hydrogenation of nitrobenzene which is catalyzed by $CuCO_3$ on silica (Sect. 2.1.4.1). To evaluate this synthesis by means of RME, a 46.3 kg mass of H_2 is introduced as the minimum amount of hydrogen needed to produce the 671.3 kg of aniline quoted in the article. A kernel RME of 0.605 is thus calculated for the process. This value decreases slightly when stoichiometry is included. Accounting for all process components further reduces the RME by 41 %. The generalized RME is much greater than that of the Suzuki reaction, which is understandable since industrial processes generally need to be much more efficient than laboratory preparations. To put this into perspective, according to a recent estimate, few synthetic schemes with more than four steps achieve general RME values above 0.15 [8]. The mass distribution for this process is divided between reagents (58 %) and reaction solvents (41 %), with catalysts and work-up/purification materials representing small quantities by comparison. This distribution further highlights the mass efficiency that is typical of industrial processes (Scheme 2.21).

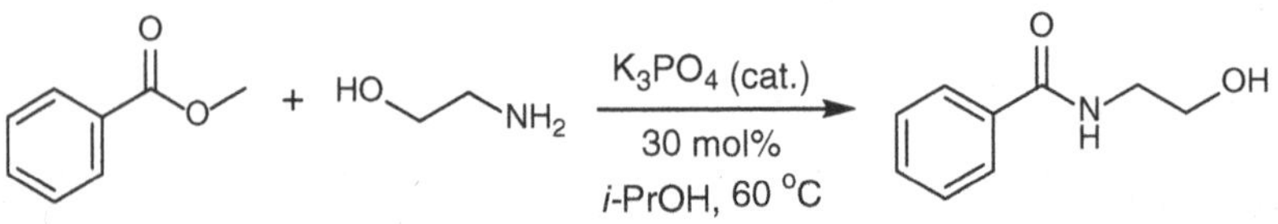

Mass: 192.2 mg   86.9 mg   188.3 mg

Moles: 1.42 mmol   1.42 mmol   1.14 mmol

GMW: 136.15   61.08   165.19

Catalyst Mass: 90 mg   Reaction Solvent Mass: 550 mg

Atom Economy = 0.84   **Yield = 0.80**   **Kernel RME = 0.674**

$$\text{Curzons RME} = \frac{188.3}{192.2 + 86.9} = 0.674$$

$$\text{Generalized RME} = \frac{188.3}{192.2 + 86.9 + 90 + 550} = 0.205$$

Scheme 2.21 Reaction mass efficiency calculations for a base-catalyzed ester amidation [64]

2.2.2.2 A Homogeneous Base-Catalyzed Amidation

Optimization of a base-catalyzed ester amidation has been described recently by Caldwell et al. [64]. The authors started with a base-mediated transesterification between an ester and an amino alcohol, which, upon rearrangement, gave the more thermodynamically stable amido alcohol product. Using catalyst and solvent screening, it was established that potassium phosphate in isopropanol had the highest product conversion. Moreover, reaction yield and the Curzons RME were used as metrics to show that the reaction achieved an average RME of 70 %, which was higher than a previously published report [54]. An example reaction (Scheme 2.23) shows that the kernel RME matches the Curzons RME meaning that no excess reagents were used.

In addition, incorporating the solvent and catalyst masses decreases the RME by 70 %. Since the catalyst represents 10 % of the entire mass distribution for this reaction, exploring catalyst recycling in future studies might result in an improved overall RME.

2.2.2.3 Biocatalysis and the Synthesis of 7-ACA

7-Aminocephalosporanic acid (7-ACA) constitutes a crucial precursor to many important semi-synthetic antibiotics including cephalosporins. In 2008, global 7-ACA production exceeded 6,000 tonnes with a market value of over US$400 million [65]. 7-ACA has been traditionally made by a four-step chemical route first developed in

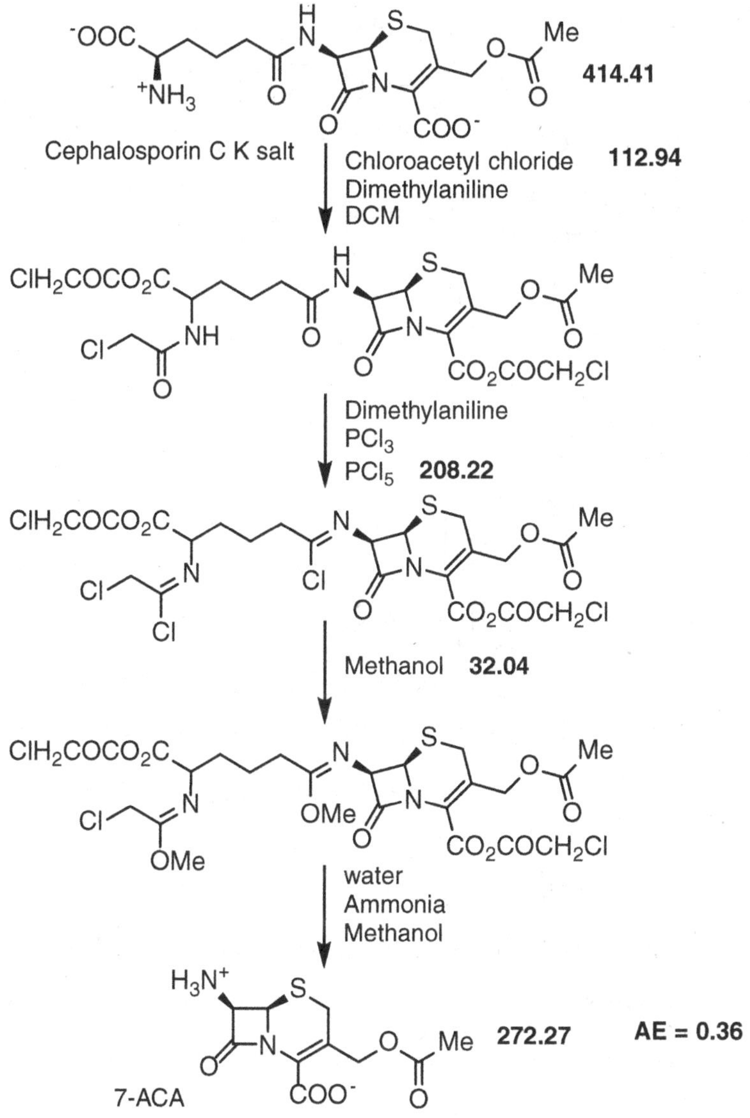

Scheme 2.22 Atom economy of the chemical route to 7-ACA [65]

1980 (Scheme 2.22) [66]. The process starts with acyl chloride protection of the amine and carboxylic acid groups of the potassium salt of cephalosporin C forming a mixed anhydride. This is then treated with phosphorus pentachloride to make an imodyl chloride which undergoes enol ether formation when treated with methanol. The resulting imodyl ether then hydrolyzes with water to give 7-ACA as the final product with an atom economy of 36 %.

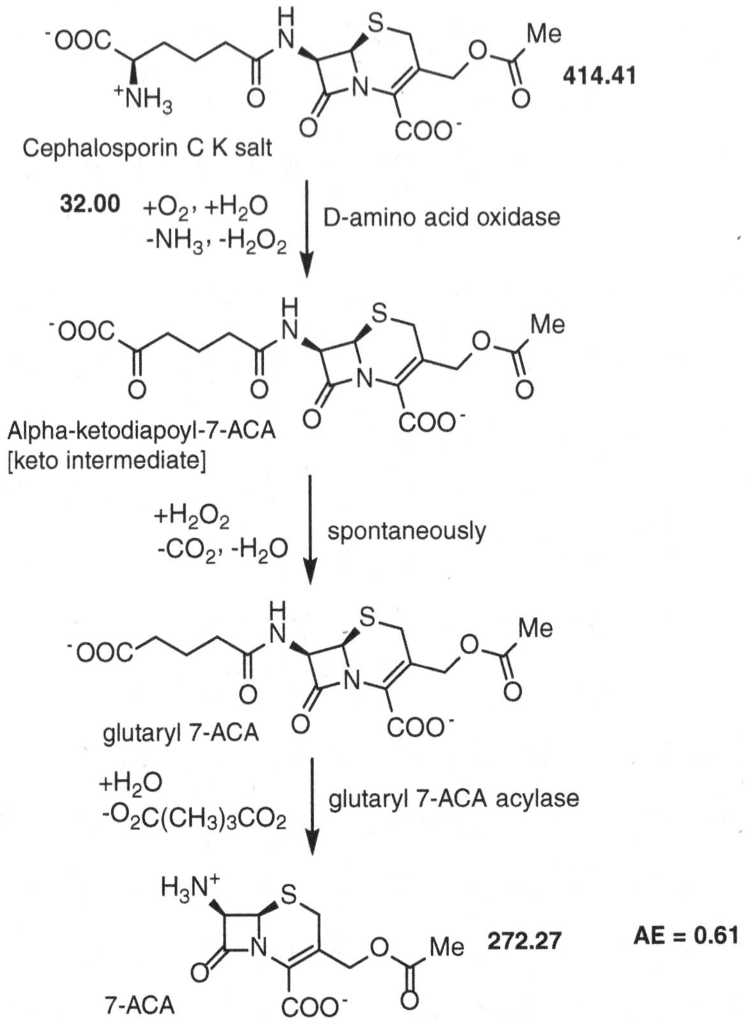

Scheme 2.23 Atom economy of the biocatalytic route to 7-ACA [65]

In comparison, the biocatalytic synthesis (Scheme 2.23) proceeds with an atom economy of 61 %. This route has completely replaced the chemical process, largely due to its 90-fold reduction in overall waste and seven-fold reduction in solvent emissions [65, 67]. Starting with a solution of the cephalosporin C potassium salt stirred with immobilized D-amino acid oxidase (DAO), reaction with oxygen gas (added via compressed air) produces a keto intermediate and hydrogen peroxide as byproduct. These then react spontaneously to form glutaryl 7-ACA which is separated from DAO (recycled) and stirred in the presence of glutaryl 7-ACA acylase (GAC) to give 7-ACA.

Table 2.1 Efficiency metrics values comparing the chemical and biocatalytic synthesis of 7-ACA calculated using RME formulas and information provided in reference [65]

Efficiency metrics	Chemical route	Biocatalysis route
Atom economy	0.355	0.610
Molar yield	0.750	0.670
Actual yield (kg)	1	1
Mass of raw materials excluding water (kg)	81	44
Reactant mass (kg)	7	3
Kernel RME	0.2663	0.4087
Curzons RME	0.1429	0.3333
Generalized RME excluding water	0.0123	0.0227
E factor including water	93	172
Generalized RME including water = $(\text{E factor} + 1)^{-1}$	0.0106	0.0058
Solvent mass excluding water (kg)	74	41
Mass of water used (kg)	13	129

A 2008 study by Henderson et al. compared these two processes according to several criteria including green metrics [65]. Combining the results of the analysis, the atom economies and several simple equations relating RME with the E factor and process mass intensity (Chap. 3), one can determine the relevant reaction mass efficiency values for these two processes (Table 2.1). Surveying these values shows that the biocatalysis route is more efficient in terms of material use when water is excluded from the measurements. Interestingly, the inclusion of water reduces the generalized RME of the biocatalysis route by 75 %.

The kernel and Curzons RMEs for this route are 53 and 133 % respectively higher than for the chemical route, clearly indicating that the biocatalysis process is more efficient. Future research should therefore aim to reduce the amount water used in the process. The mass distribution profiles for the two routes are illustrated in Fig. 2.6.

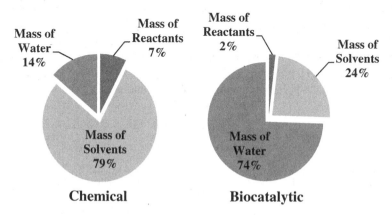

Fig. 2.6 Mass distribution profiles for the chemical and biocatalytic routes to 7-ACA

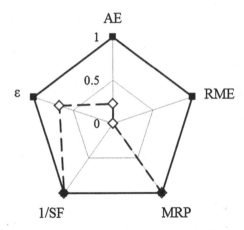

Fig. 2.7 Radial pentagon representing RME values for a Suzuki reaction

2.2.3 Future Directions

With multiple process parameters being tracked by reaction mass efficiency, organizing all the information in an intelligible manner may appear daunting. To aid with this problem, Andraos has proposed the visual model of a radial pentagon [8, 59]. Essentially, a pentagon is used to assign key reaction mass efficiency terms (atom economy, yield, RME, SF, MRP) at each corner. Assigning the edges a value of 1 and the center a value of 0 allows tracking of all five terms simultaneously along the contour of the pentagon. Points are then drawn where appropriate, and connecting adjacent points provides an immediate overall picture of how close (or far) from ideality the process is (the ideal process being represented by the perimeter of the pentagon). With this visual aid it is possible to quickly determine which terms contribute to a low RME. For example, Fig. 2.7 shows the pentagon analysis of the Suzuki reaction discussed in Sect. 2.2.1.2. One of the virtues of this approach is that it is easily extended when one introduces additional metrics with values ranging between 0 and 1.

Along with these tools, the work of Andraos includes creating process cost and energy models [68], defining environmental impact parameters on the basis of a radial polygon approach [61], predicting the intrinsic greenness of reactions given appropriately chosen thresholds [69], and creating a database of intrinsically green reactions [70].

References

1. Trost BM (1991) The atom economy—a search for synthetic efficiency. Science 254:1471–1477. doi:10.1126/science.1962206
2. Trost BM (1995) Atom economy—a challenge for organic synthesis: homogeneous catalysis leads the way. Angew Int Ed Engl 34:259–281. doi:10.1002/anie.199502591

3. Trost BM (2012) Atom economy: a challenge for enhanced synthetic efficiency. In: Li CJ (ed) Handbook of green chemistry volume 7: green synthesis. Wiley-VCH Verlag GmbH & Co. KGaA, Weinheim

4. Sheldon RA (2012) Fundamentals of green chemistry: efficiency in reaction design. Chem Soc Rev 41:1437–1451. doi:10.1039/c1cs15219j

5. Moores A (2009) Atom Economy—principles and some examples. In: Crabtree RH (ed) Handbook of green chemistry volume 1: homogeneous catalysis. Wiley-VCH Verlag GmbH & Co. KGaA, Weinheim

6. Constable DJC, Curzons AD, Cunningham VL (2002) Metrics to "green" chemistry—which are the best? Green Chem 4:521–527. doi:10.1039/b206169b

7. Constable DJC, Jimenez-Gonzalez CC (2012) Evaluating the greenness of synthesis. In: Li CJ (ed) Handbook of green chemistry volume 7: green synthesis. Wiley-VCH Verlag GmbH & Co. KGaA, Weinheim

8. Andraos J (2009) Application of green metrics analysis to chemical reactions and synthesis plans. In: Lapkin A, Constable DJC (eds) Green chemistry metrics: measuring and monitoring sustainable processes. Wiley-Blackwell, Chicester

9. Mayo DW, Pike RM, Forbes DC (2013) Microscale organic laboratory with multistep and multiscale syntheses, 6th edn. Wiley, Hoboken, pp 421–427

10. McMurry J (2012) Organic chemistry, 8th edn. Brooks/Cole, New York, pp 319–320

11. McMurry J (2012) Organic chemistry, 8th edn. Brooks/Cole, New York, pp 283–284, 289

12. Wang Z (2009) Comprehensive organic name reactions and reagents. Wiley, Hoboken, pp 2594–2599

13. VanRheenen V, Kelly RC, Cha DY (1976) An improved catalytic OsO$_4$ oxidation of olefins to cis-1,2-glycols using tertiary amine oxides as the oxidant. Tet Lett 17:1973–1976. doi:10.1016/S0040-4039(00)78093-2

14. Maurya RA, Kapure JS, Adiyala PR, Srikanth PS, Chandrasekhar D, Kamal A (2013) Catalyst-free stereoselective cyclopropanation of electron deficient alkenes with ethyl diazoacetate. RSC Adv 3:15600–15603. doi:10.1039/c3ra42374c

15. Kumagai N (2011) Development of atom-economical catalytic asymmetric reactions under proton transfer conditions: construction of tetrasubstituted stereogenic centers and their application to therapeutics. Chem Pharm Bull 59:1–22. doi:10.1248/cpb.59.1

16. McMurry J (2012) Organic chemistry, 8th edn. Brooks/Cole, New York, p 355

17. Maity AK, Chatterjee PN, Roy S (2013) Multimetallic Ir-Sn$_3$-catalyzed substitution reaction of π-activated alcohols with carbon and heteroatom nucleophiles. Tetrahedron 69:942–956. doi:10.1016/j.tet.2012.10.086

18. Ohshima T, Mashima K (2012) Platinum-catalyzed direct amination of allylic alcohols. J Synth Org Chem Jpn 70:1145–1156

19. Rothenberg G (2008) Catalysis: concepts and green applications. Wiley-VCH Verlag, New York, pp 4–28

20. Ross JRH (2012) Heterogeneous catalysis: fundamentals and applications. Elsevier, Amsterdam

21. Weissermel K, Arpe H-J (1997) Industrial organic chemistry, 3rd edn. Wiley-VCH, Weinheim, pp 143–144

22. Grant S, Freer AA, Winfield JM, Gray C, Lennon D (2005) Introducing undergraduates to green chemistry: an interactive teaching exercise. Green Chem 7:121–128. doi:10.1039/b412664e

23. Weissermel K, Arpe H-J (1997) Industrial organic chemistry, 3rd. edn. Wiley-VCH, Weinheim, pp 373–375

24. Mercer SM, Andraos J, Jessop PG (2012) Choosing the greenest synthesis: a multivariate metric green chemistry exercise. J Chem Educ 89:215–220. doi:10.1021/ed200249v

25. Constable DJC, Dunn PJ, Hayler JD, Humphrey GR, Leazer JL Jr, Linderman RJ, Lorenz KL, Manley J, Pearlman BA, Wells A, Zaks A, Zhang T (2007) Key green chemistry research areas—a perspective from pharmaceutical manufacturers. Green Chem 9:411–420. doi:10.1039/b703488c

26. Valeur E, Bradley M (2009) Amide bond formation: beyond the myth of coupling reagents. Chem Soc Rev 38:606–631. doi:10.1039/b701677h
27. Monks BM, Whiting A (2013) Direct amide formation avoiding poor atom economy reagents. In: Dunn PJ, Hii KK, Krische MJ, Williams MT (eds) Sustainable catalysis: challenges and practices for the pharmaceutical and fine chemical industries, 1st edn. Wiley, Hoboken
28. Comerford JW, Clark JH, Macquarrie DJ, Breeden SW (2009) Clean, reusable and low cost heterogeneous catalyst for amide synthesis. Chem Commun 2562–2564. doi:10.1039/b901581g
29. Yadav GD, Kadam AA (2012) Atom-efficient benzoin condensation in liquid–liquid system using quaternary ammonium salts: pseudo-phase transfer catalysis. Org Process Res Dev 16:755–763. doi:10.1021/op300027j
30. Goldman AS, Roy AH, Huang Z, Ahuja R, Schinski W, Brookhart M (2006) Catalytic alkane metathesis by tandem alkane dehydrogenation-olefin metathesis. Science 312:257–261. doi:10.1126/science.1123787
31. Foley NA, Lee JP, Ke Z, Gunnoe TB, Cundari TR (2009) Ru(II) catalysts supported by hydridotris(pyrazolyl)borate for the hydroarylation of olefins: reaction scope, mechanistic studies and guides for the development of improved catalysts. Acc Chem Res 42:585–597. doi:10.1021/ar800183j
32. Crabtree RH (2009) The organometallic chemistry of the transition metals, 5th edn. Wiley, Hoboken, pp 248–249
33. Reference 19, pp. 100–102
34. Allen LJ, Crabtree RH (2010) Green alcohol couplings without transition metal catalysts: base-mediated β-alkylation of alcohols in aerobic conditions. Green Chem 12:1362–1364. doi:10.1039/c0gc00079e
35. Gnanamgari D, Leung CH, Schley ND, Hilton ST, Crabtree RH (2008) Alcohol cross-coupling reactions catalyzed by Ru and Ir terpyridine complexes. Org Biomol Chem 6:4442–4445. doi:10.1039/b815547j
36. Gnanamgari D, Sauer ELO, Schley ND, Butler C, Incarvito CD, Crabtree RH (2009) Iridium and ruthenium complexes with chelating N-heterocyclic carbenes: efficient catalysts for transfer hydrogenation, β-alkylation of alcohols, and N-alkylation of amines. Organometallics 28:321–325. doi:10.1021/om800821q
37. Wang Z (2009) Comprehensive organic name reactions and reagents. Wiley, Hoboken, 2088–2091
38. Berliner MA, Dubant SPA, Makowski T, Ng K, Sitter B, Wager C, Zhang Y (2011) Use of an iridium-catalyzed redox-neutral alcohol-amine coupling on kilogram scale for the synthesis of a glyT1 inhibitor. Org Process Res Dev 15:1052–1062. doi:10.1021/op200174k
39. Straathof AJJ, Panke S, Schmid A (2002) The production of fine chemicals by biotransformations. Curr Opin Biotechnol 13:548–556. doi:10.1016/S0958-1669(02)00360-9
40. Parmar A, Kumar H, Marwaha SS, Kennedy JF (2000) Advances in enzymatic transformation of penicillins to 6-aminopenicillanic acid (6-APA). Biotechnol Adv 18:289–301. doi:10.1016/S0734-9750(00)00039-2
41. Powell KA, Ramer SW, del Cardayre SB, Stemmer WPC, Tobin MB, Longchamp PF, Huisman GW (2001) Directed evolution and biocatalysis. Angew Chem Int Ed 40:3948–3959. doi:10.1002/1521-3773(20020201)41:3<382:AID-ANIE2222382>3.0.CO;2-S
42. Weissenburger HWO, van der Hoeven MG (1970) An efficient nonenzymatic conversion of benzylpenicillin to 6-aminopenicillanic acid. Recl Trav Chim Pays Bas 89:1081–1084. doi:10.1002/recl.19700891011
43. Wegman MA, Janssen MHA, van Rantwijk F, Sheldon RA (2001) Towards biocatalytic synthesis of β-lactam antibiotics. Adv Synth Catal 343:559–576. doi:10.1002/1615-4169(200108)343:6/7<559:AID-ADSC559>3.0.CO;2-Z
44. Sheldon RA, Arends IWCE, Hanefeld U (2007) Green chemistry and catalysis. Wiley-VCH Verlag, Weinheim, pp 29–34
45. Dunn PJ (2012) The importance of green chemistry in process research and development. Chem Soc Rev 41:1452–1461. doi:10.1039/c1cs15041c

46. Sime JT (1999) Applications of biocatalysis to industrial processes. J Chem Educ 76:1658–1661. doi:10.1021/ed076p1658

47. Schoemaker HE, Mink D, Wubbolts MG (2003) Dispelling the myths—biocatalysis in industrial synthesis. Science 299:1694–1697. doi:10.1126/science.1079237

48. McKinney RJ (1985) Kinetic control in catalytic olefin isomerization. An explanation for the apparent contrathermodynamic isomerization of 3-pentenenitrile. Organometallics 4:1142–1143. doi:10.1021/om00125a038

49. Tolman CA (1986) Steric and electronic effects in olefin hydrocyanation at du Pont. J Chem Educ 63:199–201. doi:10.1021/ed063p199

50. Kulkarni A, Torok B (2010) Microwave-assisted multicomponent domino cyclization-aromatization: an efficient approach for the synthesis of substituted quinolines. Green Chem 12:875–878. doi:10.1039/c001076f

51. Clark H (1999) Green chemistry: challenges and opportunities. Green Chem 1:1–8. doi:10.1039/A807961G

52. Steinbach A, Winkenbach R (2000) Choose processes for their productivity. Chem Eng 107:94–104

53. Eissen M, Metzger JO (2002) Environmental performance metrics for daily use in synthetic chemistry. Chem Eur J 8:3580–3585. doi:10.1002/1521-3765(20020816)8:16<3580:AID-CHEM3580>3.0.CO;2-J

54. Curzons AD, Constable DJC, Mortimer DN, Cunningham VL (2001) So you think your process is green, how do you know?—Using principles of sustainability to determine what is green—a corporate perspective. Green Chem 3:1–6. doi:10.1039/b007871i

55. Martins MAP, Beck PH, Buriol L, Frizzo CP, Moreira DN, Marzari MRB, Zanatta M, Machado P, Zanatta N, Bonacorso HG (2013) Evaluation of the synthesis of 1-(pentafluorophenyl)-4,5-dihydro-1H-pyrazoles using green metrics. Monatsh Chem 144:1043–1050. doi:10.1007/s00706-013-0930-x

56. Stark A, Ott D, Kralisch D, Kreisel G, Ondruschka B (2010) Ionic liquids and green chemistry: a lab experiment. J Chem Educ 87:196–201. doi:10.1021/ed8000396

57. Andraos J (2005) Unification of reaction metrics for green chemistry: applications to reaction analysis. Org Process Res Dev 9:149–163. doi:10.1021/op049803n

58. Andraos J (2005) Unification of reaction metrics for green chemistry II: evaluation of named organic reactions and application to reaction discovery. Org Process Res Dev 9:404–431. doi:10.1021/op050014v

59. Andraos J, Sayed M (2007) On the use of "green" metrics in the undergraduate organic chemistry lecture and lab to assess the mass efficiency of organic reactions. J Chem Educ 84:1004–1010. doi:10.1021/ed084p1004

60. Andraos J (2009) Global green chemistry metrics analysis algorithm and spreadsheets: evaluation of the material efficiency performances of synthesis plans for oseltamivir phosphate (Tamiflu) as a test case. Org Process Res Dev 13:161–185. doi:10.1021/op800157z

61. Andraos J (2012) Inclusion of environmental impact parameters in radial pentagon material efficiency metrics analysis: using benign indices as a step towards a complete assessment of "greenness" for chemical reactions and synthesis plans. Org Process Res Dev 16:1482–1506. doi:10.1021/op3001405

62. Dicks AP, Batey RA (2013) ConfChem conference on educating the next generation: green and sustainable chemistry—greening the organic curriculum: development of an undergraduate catalytic chemistry course. J Chem Educ 90:519–520. doi:10.1021/ed2004998

63. Sakurai H, Tsukuda T, Hirao T (2002) Pd/C as a reusable catalyst for the coupling reaction of halophenols and arylboronic acids in aqueous media. J Org Chem 67:2721–2722. doi:10.1021/jo016342k

64. Caldwell N, Jamieson C, Simpson I, Watson AJB (2013) Development of a sustainable catalytic ester amidation process. ACS Sustainable Chem Eng 1:1339–1344. doi:10.1021/sc400204g

65. Henderson RK, Jimenez-Gonzalez C, Preston C, Constable DJC, Woodley JM (2008) EHS & LCA assessment for 7-ACA synthesis A case study for comparing biocatalytic and chemical synthesis. Ind Biotechnol 4:180–192. doi:10.1089/ind.2008.4.180
66. Ascher G (1980) U.S. Patent 4322526
67. Bayer T (2004) 7-Aminocephalosporanic acid—chemical versus enzymatic production process. In: Blaser HU, Schmidt E (eds) Asymmetric catalysis on industrial scale: challenges, approaches and solutions. Wiley-VCH Verlag GmbH & Co, KGaA, Weinheim
68. Andraos J (2006) On using tree analysis to quantify the material, input energy, and cost throughput efficiencies of simple and complex synthesis plans and networks: towards a blueprint for quantitative total synthesis and green chemistry. Org Process Res Dev 10:212–240. doi:10.1021/op0501904
69. Andraos J (2013) On the probability that ring-forming multicomponent reactions are intrinsically green: setting thresholds for intrinsic greenness based on design strategy and experimental reaction performance. ACS Sustain Chem Eng 1:496–512. doi:10.1021/sc3001614
70. Andraos J (2012) The algebra of organic synthesis: green metrics. CRC Press, Taylor and Francis Group, Boca Raton

Chapter 3
The E Factor and Process Mass Intensity

Abstract The environmental (E) factor and process mass intensity (PMI) metrics are introduced and thoroughly analyzed. As indispensable green metrics widely applied throughout the chemical industry, the E factor and PMI are calculated for numerous industrial processes throughout the chapter. A perspective on waste in the context of academic research, industrial synthesis and reactivity within alternative reaction media highlights the importance of material recovery, in particular with regard to reaction solvents. The section on catalysis further expands on the question of waste reduction by considering several important points. Advantages of heterogeneous catalysis which include catalyst recycling and simple product isolation and purification are described. Issues and potential solutions encountered with homogeneous catalysts and potential solutions are also discussed. Finally, the biocatalytic synthesis of pregabalin sheds light on the notions of solvent recovery and water intensity. Limitations of the E factor (which include failure to address the nature of the waste produced) provide for an introduction to process mass intensity. After explaining the simple relationship between PMI and E factor, the chapter turns to the benefits of PMI as a more robust front-end approach for evaluating the material efficiency of a process. This idea is captured by considering the biocatalytic synthesis of Singulair.

Keywords E factor · Process mass intensity · Alternative reaction media · Waste · Heterogeneous catalysis · Homogeneous catalysis · Biocatalysis · Solvent recovery · Pregabalin · Singulair

3.1 The E Factor

3.1.1 History and Development

As the chemical industry started to witness increased environmental pressures in the early 1990s, Roger Sheldon proposed the environmental (E) factor as a metric for quantifying the amount of waste produced in a chemical process. With waste

A.P. Dicks and A. Hent, *Green Chemistry Metrics*,
SpringerBriefs in Green Chemistry for Sustainability,
DOI 10.1007/978-3-319-10500-0_3

Table 3.1 E factor estimates for different chemical industries based on Sheldon's original findings [1]

Industry sector	Annual production (tonnes)	E Factor
Oil refining	10^6–10^8	<0.1
Bulk chemicals	10^4–10^6	1–5
Fine chemistry	10^2–10^4	5–50
Pharmaceuticals	10^1–10^3	25–100

defined as "anything that is not the desired product" [1], in the years following its conception, the E factor metric has contributed to significant industrial waste reduction [2–5]. In tandem with atom economy, the E factor has provided the global view of synthetic efficiency. Typical industrial E factors first published by Sheldon (Table 3.1) [1] have enhanced understanding about the problem of waste and enabled various solutions to address it. Modern techniques are now starting to show potential in bypassing certain synthetic limitations once perceived in Sheldon's original conclusions. In recent years, efforts toward waste reduction have enabled many innovations such that the majority of process chemists today consider determining an E factor as essential to process development [6].

3.1.2 Intrinsic and Global E Factors

The E factor approach to waste complements the atom economy metric because it offers both intrinsic and global evaluations of a synthetic process. The intrinsic method is one which Andraos calls the environmental impact factor based on molecular weight (E_{mw}) [7]. This metric is calculated as the ratio of the molecular weight of all by-products divided by the molecular weight of the desired product. The E factor based on mass (E factor) offers a global perspective and is calculated as the ratio of the total mass of all waste to the mass of the desired product [1]. Figure 3.1 outlines the calculations for both metrics in the context of a balanced stoichiometric reaction.

$$A + B \xrightarrow[\substack{solvents \\ catalysts}]{reagents} C + D + waste$$

Mass:	m_1	m_2	m_3	m_4	m_5
Moles:	n_1	n_1	n_2	n_2	-
GMW:	MW_1	MW_2	MW_3	MW_4	-

$$E_{mw} = \frac{MW_4}{MW_3} \qquad E\ factor = \frac{(m_1 + m_2 + m_4 + m_5) - m_3}{m_3}$$

Fig. 3.1 Generic addition reaction where an equal molar amount of A and B react together to produce n_2 mol of a desired product C, a byproduct D and unrecovered waste

$$\text{Atom Economy} = \frac{MW_3}{MW_1 + MW_2} = \frac{MW_3}{MW_3 + MW_4}$$

$$\frac{1}{\text{Atom Economy}} = \frac{MW_3 + MW_4}{MW_3} = 1 + \frac{MW_4}{MW_3} = 1 + E_{mw}$$

$$\text{Hence:} \quad \text{Atom Economy} = \frac{1}{1 + E_{mw}}$$

Fig. 3.2 Derivation of the relationship between E_{mw} and atom economy for a balanced stoichiometric reaction (i.e. $MW_1 + MW_2 = MW_3 + MW_4$)

From the conservation of mass law it is possible to derive simple mathematical relationships between the E factor metrics and the metrics discussed in Chap. 2. These are inverse relationships since the atom economy and reaction mass efficiency metrics account for material input whereas the E factor metrics account for material output. As an illustration, Fig. 3.2 presents the derivation of the relationship between atom economy and E_{mw} for the stoichiometric reaction in Fig. 3.1. The same relationship holds for reactions under non-stoichiometric conditions.

A similar equation relating the global E factor and the generalized RME (Sect. 2. 2.1.2) was derived by Andraos [7] and will be explained in Sect. 3.2.2. Applying the E factor analysis to the familiar Suzuki reaction (Scheme 3.1) shows that copious amounts of waste are produced both intrinsically and globally. In terms of waste, this Suzuki reaction actually exceeds the typical pharmaceutical industry range (Table 3.1). Similar examples highlight how little attention is often attributed to the problem of waste in an academic setting (Sect. 3.1.3). Lastly, it is important to note that the È factor, just like the generalized RME, accounts for product yield, stoichiometry, catalysts, solvents, and other auxiliary masses used.

The distinction between intrinsic and global waste is most evident when comparing the E_{mw} and E factor metrics for an antiquated phloroglucinol process [4]. As a useful intermediate for the preparation of pharmaceuticals, dyestuffs, perfumes, polymer additives and adhesives, phloroglucinol was traditionally made via a three-step chemical route with an overall yield of >90 % (Scheme 3.2) [5]. Starting with the potassium dichromate oxidation of 2,4,6-trinitrotoluene (TNT), the process follows Béchamp reduction with iron and hydrochloric acid and ends with hydrolysis from heating [4]. On the basis of the product yield this process was once regarded as being efficient.

However, the intrinsic chemistry reveals a significant formation of reaction by-products ($E_{mw} = 17.1$). This value translates to an atom economy of approximately 6 %. Note that even an E_{mw} value of 1 implies an atom economy of 50 %. The E_{mw} in general is a significant measure of efficiency only if the other elements of the process are 100 % efficient. The traditional phloroglucinol process produces approximately 40 kg of solid waste containing $Cr_2(SO_4)_3$, NH_4Cl, $FeCl_2$ and $KHSO_4$ per kg of final product [5]. The E factor is therefore 40. The reason behind

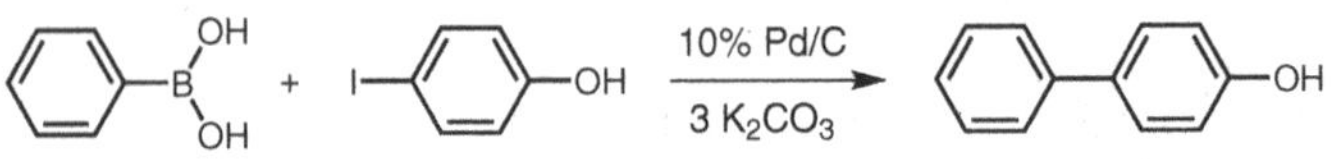

0.122 g   0.220 g   0.415 g   0.115 g

Catalyst Mass: 0.003 g   Reaction Solvent Mass: 11 g

Work-up and Purification Material Mass: 38.1 g

Atom Economy = 0.225   **Yield = 0.675**   **Generalized RME = 0.0023**

By-products: I^- HCO_3^- CO_3^{2-} $6\,K^+$ $^-O_2CO-B$

GMW: 126.91   61.02   60.01   234.59   104.83

$$E_{mw} = \frac{126.91 + 61.02 + 60.01 + 234.59 + 104.83}{170.21} = 3.45$$

$$E\ \text{factor} = \frac{0.122 + 0.220 + 0.415 + 0.003 + 11 + 38.1 - 0.115}{0.115} = 433$$

$$AE = \frac{1}{1 + 3.45} = 0.225 \qquad gRME = \frac{1}{1 + 433} = 0.0023$$

Scheme 3.1 Metrics analysis for a Suzuki reaction undertaken as an undergraduate experiment [95]

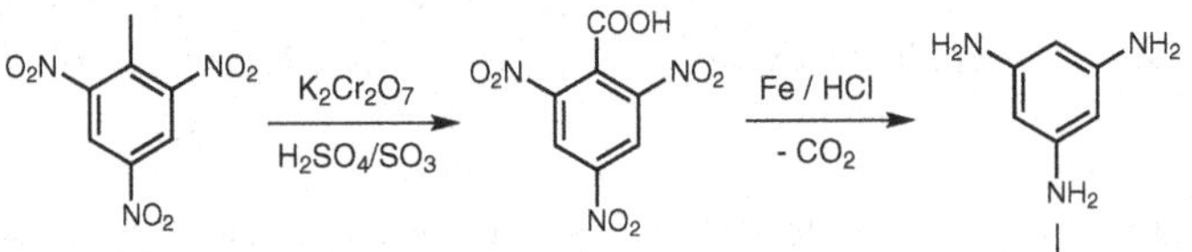

Balanced Chemical Equation:

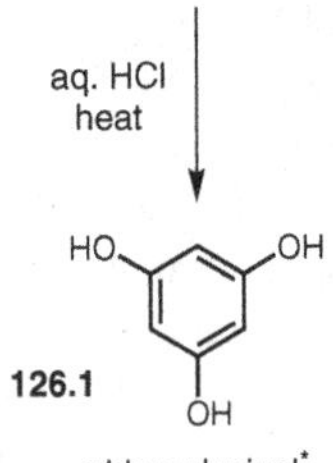

Input: $C_7H_5N_3O_6 + K_2Cr_2O_7 + 5\ H_2SO_4 + 9\ Fe + 21\ HCl$

Output: $C_6H_6O_3{}^{\cdot} + Cr_2(SO_4)_3 + 2\ KHSO_4 + 9\ FeCl_2$
         $+ 3\ NH_4Cl + CO_2 + 8\ H_2O$

126.1

phloroglucinol$^{\cdot}$

By-products: $Cr_2(SO_4)_3$ $2\ KHSO_4$ $9\ FeCl_2$ $3\ NH_4Cl$ CO_2 $8\ H_2O$

GMW: 392.2   272.3   1140.7   160.5   44.0   144.1

$$E_{mw} = \frac{392.2 + 272.3 + 1140.7 + 160.5 + 44.0 + 144.1}{126.1} = 17.1 \qquad E\ \text{factor} = 40$$

Scheme 3.2 Intrinsic and global E factors for the phloroglucinol chemical process [4, 5]

the waste involves use of excess oxidizing and reducing agents as well as an excess of sulfuric acid, which has to be neutralized with base [4, 5]. With time the cost of treating and handling the waste started to reach the selling point of the product [8], at which point the chemical route had to be phased out. Today, phloroglucinol can be made more efficiently via biosynthesis [9], e.g. by using gene expression in model organisms [10, 11].

Since any and all waste is undesirable, the ideal E factor is zero. For intrinsic waste, calculating atom economy is often simpler because by-products do not have to be determined. When calculating an E factor it is important to keep the entire process in mind. A recent report discussing waste reduction in the synthesis of fine chemicals calculates process E factors through simple summation of the E factors of individual chemical steps [12]. It should be stressed that the E factor metrics are *not* additive [7]. In order to determine E factors for a multi-step synthesis one must add together all the unrecovered waste produced in every individual step and divide it by the mass or molecular weight of the final product of the entire synthesis. Adding individual E factors together results in arbitrarily lower values. Discussion of the proper ways to interpret and report E factors will follow in subsequent sections.

3.1.3 Perspective on Waste in Academia and Industry

Determining E factors can sometimes overshadow other important aspects of waste in the context of green chemistry. In order to introduce students to the E factor one must adopt a more comprehensive analysis of waste, including its causes and the means of optimizing for reduction and elimination. As a solution, Sheldon advised the substitution of "antiquated stoichiometric methodologies with green catalytic alternatives that are more atom economic" so that the amount of inorganic salt generated by the former can be reduced [5]. Although this is an excellent proposal, students may find it too complicated when first learning about the E factor (Sect. 3.1.4). For example, recycling or replacing the catalyst in the Suzuki reaction without any solvent recovery would not improve the E factor by much, and attempting it might result in a higher cost. Instead, students should be encouraged to explore ways to improve atom economy, the possibility of replacing or recycling the reaction media (Sect. 3.1.5), and most of all the possibility of optimizing the experimental procedure provided. It should always be stressed that an experimental procedure is never "set in stone".

Two recent articles published in the *Journal of Chemical Education* highlight possible experimental solutions to the problem of waste. In 2013, Stacey et al. used the environmentally benign solvent polyethylene glycol (PEG-400) in combination with racemic proline (an organocatalyst) to carry out carbonyl condensation reactions in an undergraduate laboratory (Scheme 3.3) [13]. Due to the high boiling point of the PEG-400/proline mixture, it was possible to perform multiple experiments by simply recycling the reaction mixture using a standard rotary evaporator.

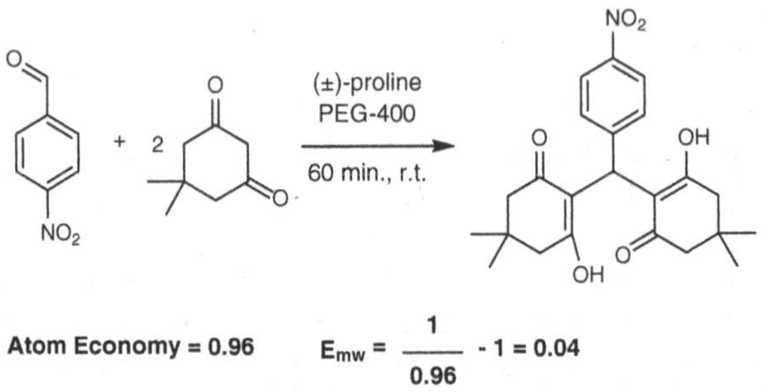
(±)-proline
PEG-400
60 min., r.t.
NO2
OH
Atom Economy = 0.96
Emw = 1/0.96 - 1 = 0.04

Scheme 3.3 One-pot carbonyl condensation reaction undertaken by students using proline as an organocatalyst and polyethylene glycol (PEG-400) as a recyclable solvent [13]

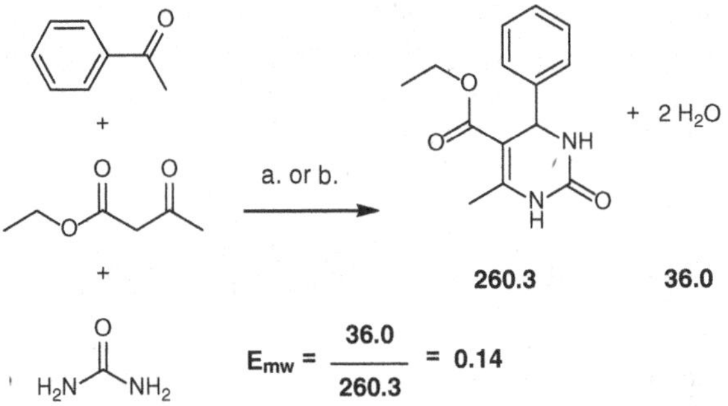
a. or b.
+ 2 H2O
NH
260.3
36.0
Emw = 36.0/260.3 = 0.14

Scheme 3.4 Traditional (**a** HCl, EtOH, heat, 90 min.) and modern (**b** $ZnCl_2$, solvent free, heat, 15 min) Biginelli reactions [13]

In another study, Aktoudianakis et al. compared a traditional synthesis with a modern solvent-free approach in the context of the Biginelli reaction (Scheme 3.4) [14]. This multicomponent reaction has an E_{mw} of 0.14 (88 % atom economy) and is ideal in an undergraduate laboratory setting, as the solvent-free reaction is completed within 15 min. Although the E factor metrics were not determined in these articles, applying them would show students the value of solvent recycling and alternative reaction media in terms of minimizing waste. For more information on waste reduction in an educational setting, the reader is referred to other excellent sources [15–18].

One waste reduction strategy often employed in the chemical industry involves a streamlined process which uses a continuous flow reactor [19]. This method uses flow tubes to steadily transfer feedstock into a reaction mixture and separate the final product as it forms. Waste reduction is therefore achieved through recycling of the reaction mixture. Reports comparing new processes to traditional batch methods with calculated E factors are becoming more frequent in the literature [20, 21]. In addition, achievements in synthetic efficiency at pharmaceutical companies show that improving E factors can lead to improvements in other areas related to process greenness [22].

Nevertheless, reporting E factors in journals is sometimes plagued by failure to distinguish between the intrinsic E_{mw} and the global E factor, as well as uncertainty regarding the kind of waste included in the calculation. For instance, Sheldon's initial position was that water should not be included in the E factor because this would produce extremely high values making meaningful comparison of processes difficult [1–3]. In recent years, owing to the significant quantity of aqueous waste generated in certain pharmaceutical processes, Sheldon's perspective has changed to suggest the inclusion of water may actually be warranted [4, 5]. Indeed, articles providing E factors with and without the inclusion of water have started to become more numerous.

3.1.4 The Solution: Catalysis

As most of the waste produced in traditional industrial processes consisted of inorganic salts, (a consequence of stoichiometric methods originally based on a "yield-only" mentality) Sheldon proposed that the solution to the problem of waste was catalysis [1]. More specifically he stated that "synthetic organic chemists would benefit enormously from the application of catalytic retrosynthetic analysis to identifying routes to a desired product" [1]. Note that waste reduction naturally implies that the best catalyst is none at all. Consequently, in terms of catalysis, it is important to emphasize material recovery without compromising reactivity or efficiency. Unfortunately, in this regard, not all forms of catalysis are created equal.

3.1.4.1 The Advantage of Heterogeneous Catalysis

Heterogeneous catalysis, whenever possible, is an established and preferred method of reducing waste [3–5]. This is because simple filtration and/or centrifugation facilitate the recovery of a heterogeneous catalyst from solution, leaving minimal impurities in the final product [3]. The palladium catalyst in the Suzuki reaction for instance can be recycled through gravity filtration (Sect. 2.2.1.2) [23]. This highly desirable quality of heterogeneous catalysis can be easily incorporated in an undergraduate laboratory experiment.

An important aspect pertaining to catalysis in general is that one should always attempt to search for improved catalytic efficiency. With regard to the E factor, the amide synthesis originally discussed in Sect. 2.1.4.1 illustrates this point (Scheme 3.5). Although the K60 silica gel catalyst used by Comerford required activation at high temperatures, the amide synthesis produced global E factors approaching 1 [24]. These values reflected 90 % solvent recovery in combination with catalyst recyclability (up to five times). When compared with traditional methods using inefficient reagents like DCC and $SOCl_2$, the new catalyst demonstrated that significant waste reduction was possible. Detailed calculations of E factors for the different methods compared are available in the original article.

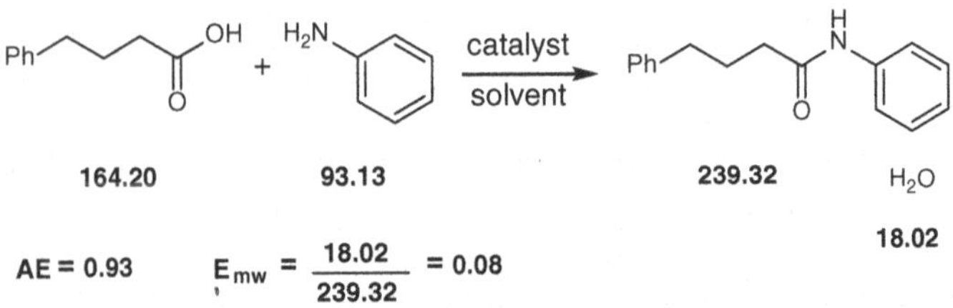

Ph
OH
O
H2N
+
catalyst
solvent
Ph
H
N
O
164.20
93.13
239.32
H2O
18.02
AE = 0.93
Emw = 18.02 / 239.32 = 0.08

Scheme 3.5 Synthesis of 4,*N*-diphenylacetamide

Progress on this reaction came in 2010, when an article appeared describing a new sulfated tungstate-catalyzed synthesis [25]. This method produced higher E factors, most likely because solvent recovery was not attempted. Nevertheless, the new heterogeneous acid catalyst did not require activation at high temperatures and produced higher yields as compared with the K60 silica catalyst. In 2012, Ghosh et al. added activated alumina balls to the list of efficient heterogeneous catalysts for the amide synthesis [26]. Using green metrics to compare their method to the Comerford synthesis, the authors showed that the new catalyst had the ability to work in solvent-free conditions [26]. Although activation at high temperature was still necessary, catalyst recovery for several cycles produced noticeably lower E factors, showing that a more efficient synthesis had been achieved. A recent article describing a new sulfated tungstate-catalyzed Ritter reaction [27] also shows how an E factor analysis can help determine which heterogeneous catalyst is the most efficient [28].

On an industrial scale, Sheldon has discussed use of titanium(IV) substituted silicate-1 (TS-1) as an efficient heterogeneous catalyst for the manufacture of caprolactam, an important precursor to nylon 6 [4]. The process starts with the TS-1 catalyzed ammoximation of cyclohexanone in NH_3–H_2O_2 followed by a Beckmann rearrangement carried out in the vapour phase over a high-silica MFI zeolite, another heterogeneous catalyst (Scheme 3.6) [29, 30]. The use of zeolites in synthesis and in green chemistry is discussed in detail elsewhere [31, 32]. The catalyzed caprolactam process produces only two molecules of water as intrinsic waste, and is essentially salt-free in a global sense. In contrast, the conventional route [33] produces almost 4.5 kg of ammonium sulfate waste per kg of caprolactam formed [4], since two of the steps require reagents in stoichiometric quantities (Scheme 3.7).

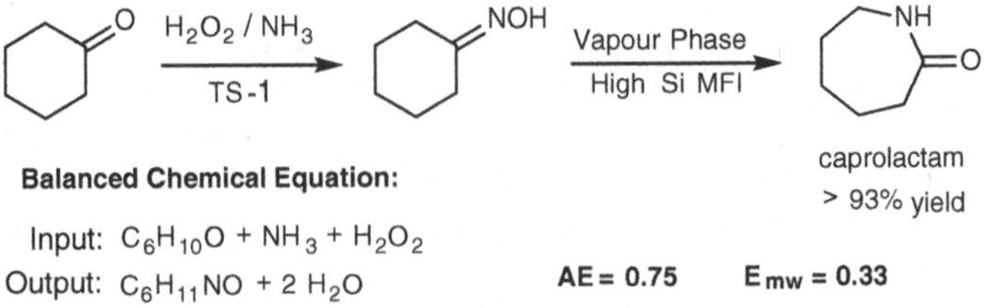

O
H2O2 / NH3
TS-1
NOH
Vapour Phase
High Si MFI
NH
O
caprolactam
> 93% yield
Balanced Chemical Equation:
Input: C6H10O + NH3 + H2O2
Output: C6H11NO + 2 H2O
AE = 0.75
Emw = 0.33

Scheme 3.6 TS-1 catalyzed process for the synthesis of caprolactam

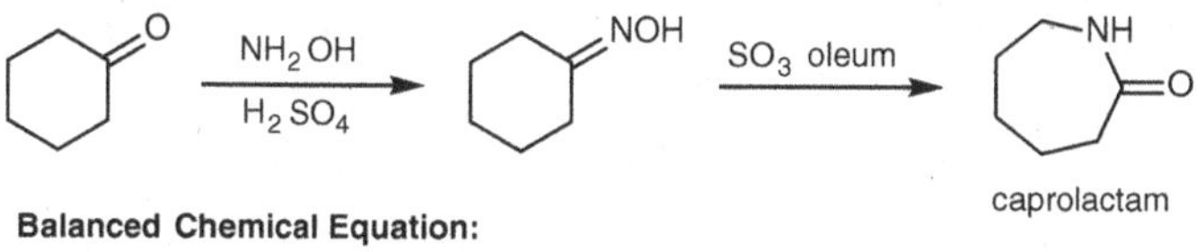

Balanced Chemical Equation:

Input: $C_6H_{10}O + 0.5 (NH_3OH)_2 SO_4 + 1.5 H_2SO_4 + 4 NH_3$

Output: $C_6H_{11}NO + 2 (NH_4)SO_4 + H_2O$

AE = 0.29 E_{mw} = 2.5

Scheme 3.7 Conventional process for the synthesis of caprolactam

Other examples of heterogeneous catalysis applied in industry have been discussed elsewhere [34]. Finally, it is known that heterogeneous catalysts are generally less active and may present heat transfer and selectivity issues when compared with homogeneous catalysts [35]. Nevertheless, their simple recovery makes them highly attractive compounds from the point of view of waste reduction.

3.1.4.2 Opportunities in Homogeneous Catalysis

Homogeneous catalysis usually provides milder conditions, more efficient heat transfer, and higher selectivity/activity when compared to heterogeneous conditions [35]. Despite these advantages, difficult catalyst recoveries and product purifications present serious drawbacks in terms of waste reduction [4]. The BHC ibuprofen process, for instance, requires a difficult separation with an expensive purification for an otherwise intrinsically efficient catalytic step (Scheme 3.8) [4]. A solution to this problem required using a biphasic environment where the catalyst dissolves in the aqueous phase while the product dissolves in the organic phase, thus enabling phase separation [36]. Methods to heterogenize homogeneous catalysts such as immobilization on organic or inorganic supports generally suffer from metal leaching, poor productivity, support degradation and irreproducible activities/selectivities, making them commercially problematic [35]. Progress on immobilization and multi-phase catalysis under continuous flow conditions has been reported [37, 38]. For a general discussion on multi-phase homogeneous catalysis, the reader is referred to reference [39]. In an industrial setting, multi-phase homogeneous catalysis is applied in the Rhône-Poulenc/Ruhrchemie process for aqueous biphasic hydroformylation [35].

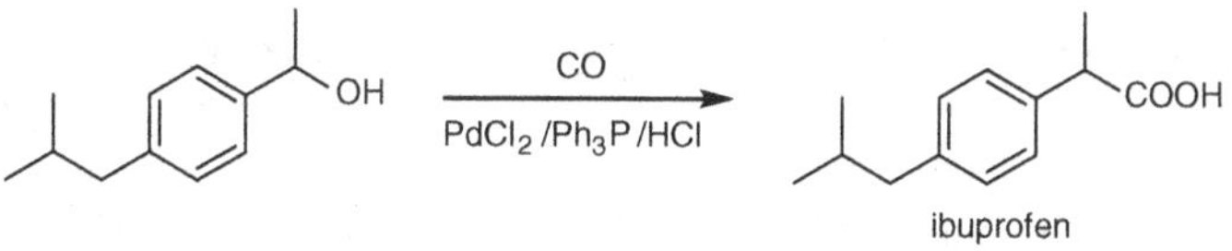

Scheme 3.8 The final synthetic step for the modern BHC ibuprofen synthesis

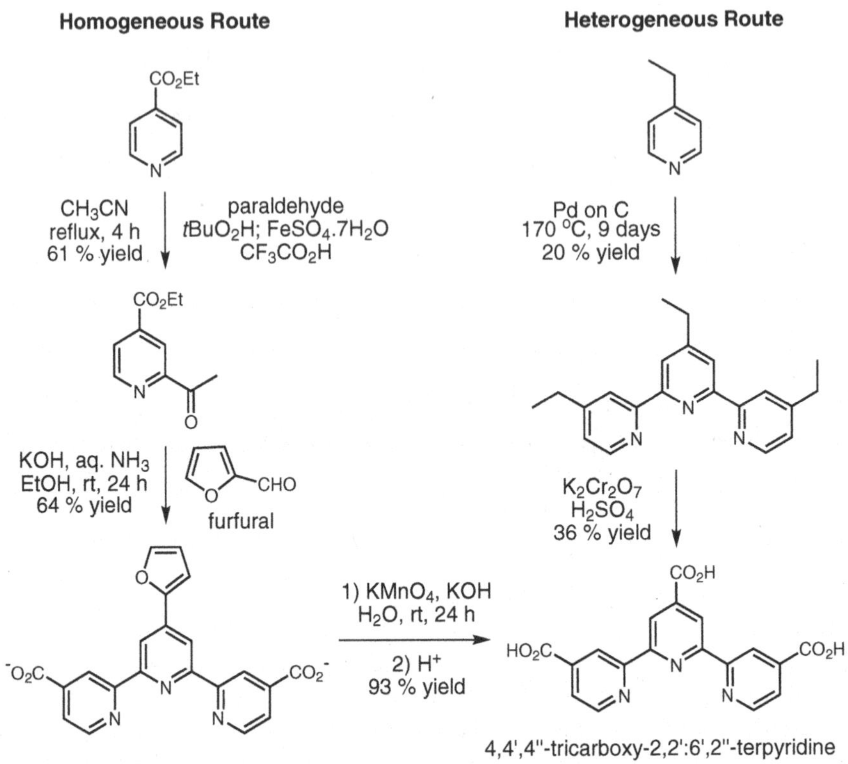

Homogeneous Route   **Heterogeneous Route**

Homogeneous Route: AE = 15%   Yield = 36%   E factor = 106.1   Cost = 6.47 $/g

Heterogeneous Route: AE = 8.2%   Yield = 7.2%   E factor = 336.4   Cost = 7.79 $/g

Scheme 3.9 Metrics analysis for two syntheses of 4,4',4''-tricarboxy-2,2':6',2''-terpyridine

Although general consensus suggests avoiding [40, 41] or at least improving homogeneous catalysis via immobilization [42, 43], a recent article showed how a novel organo-catalyzed synthesis is made more efficient than the original hetero-geneous route (Scheme 3.9) [44].

The homogeneous route to 4,4',4''-tricarboxy-2,2':6',2''-terpyridine, a precursor to complexes used in dye-sensitized solar cells, has an atom economy of 15 % vs. 8 % for the heterogeneous route. In addition to a five-fold increase in yield and a three-fold reduction in waste (E factor), the new route also achieved a lower overall cost. These results took into account catalyst and solvent recycling for the first step of the heterogeneous route. Moreover, the new synthesis used furfural, a compound easily obtained from renewable sources such as corn cobs. Despite these improvements, an E factor of 106 is still exceptionally high for a three-step syn-thesis, suggesting that future research might focus on developing a solvent recovery solution, perhaps in a biphasic environment.

3.1.4.3 Biocatalysis and the Issue of Solvent Waste

Biocatalysis was recently described as the "main green chemistry technology adopted by the fine chemicals and pharmaceutical industries" today [45]. Although enzymes typically incur a higher cost, biocatalysis is preferred for its high selectivity and mild reaction conditions which can extend the life span of bioreactors. Biocatalysis also makes use of aqueous media and biodegradable enzymes which are often extracted from renewable sources. In this context, Sheldon has highlighted the greater feasibility regarding disposal of organic waste as compared to inorganic waste [3].

By applying biocatalysis, it is also possible to significantly reduce the amount of waste generated in older chemical processes. To appreciate this it is worth revisiting the manufacture of 6-aminopenicillanic acid (6-APA) discussed in Chap. 2 (Schemes 2.16 and 2.17). Here, the traditional chemical route has an E factor of 20.4 with waste consisting of: 0.6 kg Me_3SiCl, 1.2 kg PCl_5, 1.6 kg $PhNMe_2$, 0.2 kg NH_3, 8.4 kg n-BuOH and 8.4 kg CH_2Cl_2 [46]. In contrast, the enzymatic route uses only 1–2 kg of Pen-acylase, 0.1 kg NH_3 and 2 kg H_2O for an E factor of 3.1–4.1 [47]. This waste is also significantly less hazardous. It should be noted that in the context of biocatalysis, it is crucial to include water in the E factor calculation. This is because biocatalysis methods are usually highly water-intensive processes. Applying the E factor without the inclusion of water can shift environmental burden from waste reduction to water consumption.

In a recent review, Dunn described the development of Pfizer's biocatalytic resolution of pregabalin (Fig. 3.3) from an efficiency and waste reduction perspective [48]. Pregabalin is a GABA analog used for the treatment of various nervous system disorders, anxiety and social phobia, with a market of $3.06 billion in the US in 2010 [48].

A biocatalytic route formulated in 2006 reduced the E factor of the resolution step from 86 to 17. This value takes into account significant solvent recovery by way of water treatment and organic phase separation [49]. In 2010, solvent recovery in combination with recycling of the opposite enantiomer via a new base-catalyzed epimerization further reduced the E factor to only 8 [48]. The prospects of solvent recycling in biocatalysis and industrial biotechnology in general have been reviewed elsewhere [50, 51]. It should also be noted that the E factors reported by Dunn do not include process water. This discrepancy leaves questions regarding the water intensity of the biocatalytic process unanswered.

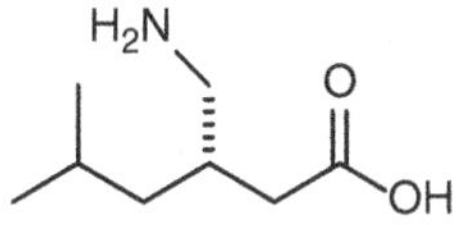

Fig. 3.3 Structure of pregabalin

3.1.5 Perspectives on Waste in Alternative Reaction Media

Development of alternative reaction media has recently accelerated after reports demonstrated that waste produced in the chemical industry is largely attributed to solvent loss [3, 4, 52, 53]. Non-traditional media like water, supercritical fluids, ionic liquids, and solvent-free conditions (SFC) provide chemists with access to faster reaction times, higher yields, renewable feedstocks, novel chemistries, and new opportunities for material recovery and recycling [3, 4, 35]. In addition, they also facilitate targeted efforts at waste reduction.

For many years, chemists believed that the only good alternative to organic-based solvents was water. Along with its low toxicity, inflammability and ubiquitous nature, water provides two important synthetic features: a high specific heat capacity which can facilitate exothermic reactions [54] and the hydrophobic effect which can enhance reactions of non-polar materials in aqueous suspensions ("on water" reactions) [55]. Polarity and easy separation of insoluble products also play important roles in synthesis, especially in the context of organometallic catalysis [56]. In a recent study, Isley et al. described novel aqueous Suzuki-Miyaura reactions with improved E factors [57]. The new approach involved coupling of N-methyliminodiacetic acid (MIDA) boronates with aryl bromides in water at room temperature (Scheme 3.10). A near ideal E factor (excluding water) and an aqueous E factor of 6.5 were quoted by the authors. When compared to the student-run Suzuki reaction (Sect. 3.1.2), these E factor values favor very well.

Like water, supercritical fluids (SCFs) have recently attracted much interest as reaction solvents [35]. Defined as compounds in a state above their critical pressure and temperature [58], SCFs are sought-after because of their non-volatility, inertness, and tunable solubility. In the case of supercritical carbon dioxide ($scCO_2$, one of the most commonly used SCFs in organic synthesis), one can control solubility by varying the solvent temperature and pressure above the supercritical region (74 bar and 31 °C) [35]. Other green qualities of $scCO_2$ include removal by depressurization, non-toxic chemical inertness, and no added greenhouse gas emissions. Besides extracting caffeine from coffee, $scCO_2$ and other SCFs have

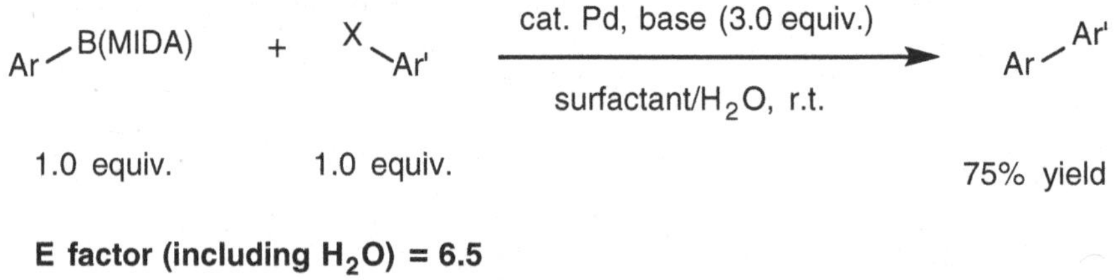

$$Ar\diagup B(MIDA) \quad + \quad X\diagdown_{Ar'} \quad \xrightarrow[\text{surfactant/}H_2O,\ r.t.]{\text{cat. Pd, base (3.0 equiv.)}} \quad Ar\diagup Ar'$$

1.0 equiv.   1.0 equiv.   75% yield

E factor (including H_2O) = 6.5

Scheme 3.10 Suzuki-Miyaura cross-couplings of N-methyliminodiacetic acid (MIDA) boronates and aryl bromides in water at room temperature [57]

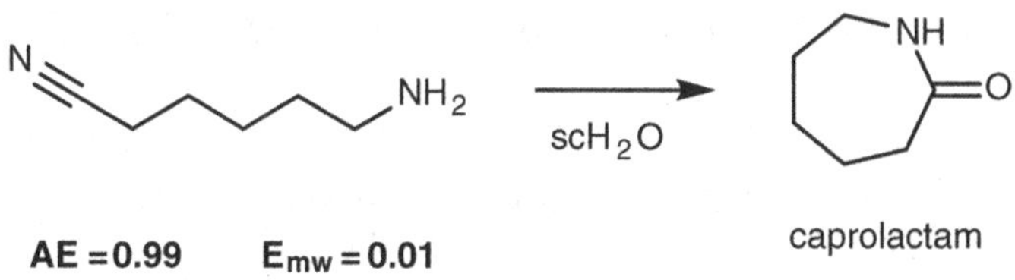

AE = 0.99   E$_{mw}$ = 0.01   caprolactam

Scheme 3.11 Synthesis of caprolactam from 6-aminocapronitrile in scH$_2$O

been applied in fields such as hydrogenation, hydroformylation and biocatalysis [35, 58–60]. For waste reduction, the production of caprolactam from 6-aminoc-apronitrile in scH$_2$O (Scheme 3.11) affords a much lower E$_{mw}$ than the TS-1 catalyzed process discussed previously (Scheme 3.6) [61]. Nevertheless, using SCFs for synthesis often necessitates equipment capable of safely handling high pressures.

Other alternative media gaining attention are easily recyclable liquid polymers (including poly(ethylene glycol), e.g. PEG-400), solvents from renewable sources (e.g. 2-methyltetrahydrofuran), switchable solvents with tunable polarity, and ionic liquids. The latter are the most commonly applied media under the umbrella of green chemistry [59]. Ionic liquids (ILs) are salt mixtures with melting points <100 °C [59]. Their popularity is largely based on their low volatility, thermal stability over a wide temperature range, and adjustable properties. However, toxicity and environmental footprint data suggest that ILs do not constitute a green technology [62, 63]. In order to change this perception, efforts have been focused on developing green assessment tools for ionic liquid synthesis [64]. Concerning waste reduction, enantioselective alcohol syntheses in ILs with significantly improved E factors were recently reported [50, 65].

Perhaps the most significant category of alternative reaction media from the perspective of waste elimination is that of solvent-free conditions (SFC). Since Constable's 2002 article which estimated that solvents constitute ca. 85 % of the mass involved in pharmaceutical manufacture [66], research into SFC reactions has greatly intensified [26, 28, 43]. For example, a thorough review describing their impact in heterocyclic synthesis appeared in 2009 [67]. Therein, Martins et al. showed that SFC reactions take less time to complete and give higher or similar yields versus the same reactions carried out with molecular solvents. Moreover, solvent-free reactions where only the synthesis step was considered had an E factor range of 0.1–4.0, which falls into the bulk chemicals category of waste production (Table 3.1). Nevertheless, none of the articles considered in the review accounted for the waste produced during product purification, which for SFC reactions may require large amounts of solvents and energy. The procedures which described product isolation had an E factor range of 24-389. This difference highlights the need for full disclosure of experimental conditions in the literature.

3.1.6 Beyond the E Factor: Innovative Synthetic Methods

The E factor metric has encouraged chemists to tackle the issue of waste for over 20 years. Striving for better environmental and economic syntheses, chemists had achieved the optimization of many processes, including those of sildenafil citrate (Viagra) [68] and sertraline [69]. Today, innovative waste reduction methods are numerous and diverse, appearing in such areas as catalysis [70, 71], catalyst recovery [72], solvent recovery [73] and optimization of work-up conditions [74]. Without the E factor metric, the field of green chemistry would not be where it currently is.

Nevertheless, while the E factor is a tangible metric used for quantifying generated waste, its failure to address the *nature* of the waste has driven others to propose qualitative assessment tools including industrial solvent selection guides [75] and the EcoScale (Chap. 4). Another important challenge involves the value of the "back-end" perspective on efficiency offered by the E factor. With increasingly complex molecules requiring multi-step syntheses, it is expected that the value of material output would drop as compared to that of material input, hence necessitating a different green metric to provide a "front-end" analysis. To this end, Sheldon offered a worthy piece of advice in his landmark 1992 article that introduced the E factor to the world: "Problems are just opportunities in disguise [1]."

3.2 Process Mass Intensity (PMI)

3.2.1 History and Motivation

In 1991, atom economy was introduced as a front-end approach for assessing reaction efficiency from an intrinsic perspective taking into account material input and determining the extent of its incorporation into a final product (Chap. 2) [76]. A corresponding metric for global efficiency did not appear until 1998 when Heinzle et al. defined "mass intensity" as the mass ratio of total input material to final product [77]. In subsequent years, this metric had been rebranded as "mass index" [78] and more recently as "process mass intensity" (PMI) in an effort to benchmark environmental acceptability standards for key pharmaceutical companies [53]. Today, despite Sheldon's insistence that the E factor is better at reflecting the goals of green chemistry [3], it has been estimated that PMI is used by 67 % of chemical companies while the E factor is used only by 48 % [79]. These statistics are unsurprising considering the philosophical and technical arguments put forward by GSK chemists in a 2011 article explaining the value of PMI [53]. From environmental (carbon footprint contribution) and economic perspectives, it is clear that much more energy and capital is spent on the material input side of API manufacturing as compared to waste treatment and handling. Consequently, using PMI as a front-end approach to determine the global efficiency of a synthesis has proven very valuable for the chemical industry.

$$PMI = \frac{\text{total M used in process (kg)}}{\text{M product (kg)}} \quad (M = mass)$$

$$= \frac{\text{total M used in process (kg) - M product (kg) + M product (kg)}}{\text{M product (kg)}}$$

$$= \frac{\text{total M of waste (kg) + M product (kg)}}{\text{M product (kg)}} = E\ factor + 1$$

Fig. 3.4 Relationship between process mass intensity and the E factor

3.2.2 Process Mass Intensity in Relation to Other Metrics

Process mass intensity is defined as the "total mass of materials used to produce a specified mass of product" where "materials include reactants, reagents, solvents used for reaction and purification, and catalysts" [53]. This definition is reminiscent of Andraos' generalized reaction mass efficiency (gRME, Sect. 2.2.1.2), only PMI is actually the inverse of gRME [7]. One can also calculate PMI by adding 1 to the E factor (Fig. 3.4). Consequently, the ideal PMI for a process where all material input ends up in the product is equal to 1. In this context, Sheldon criticized process mass intensity on the premise that an ideal value of 1 does not effectively emphasize the ideal goal of zero waste produced in a synthesis [3]. Interestingly, he also wrote that "green chemistry is primary pollution prevention rather than waste management (end-of-pipe solutions)" [4].

One of the main differences between PMI and E factor is that PMI explicitly includes water in the calculation [80]. In addition, Jimenez-Gonzalez et al. claimed that the difference of one between PMI and the E factor actually represents "saleable product", which means that PMI directly emphasizes a company's revenue stream [53]. This, they argued, promotes a productivity-oriented mentality which contributes to "reinventing 'business-as-usual', especially in the broader context of the supply chain" [53]. Moreover, life cycle data obtained from GSK showed that a much greater carbon footprint is associated with the supply chain (94 %) than with waste treatment (6 %) [53]. Since PMI measures input efficiency, it can complement a life cycle analysis much better than the E factor could. Lastly, Jimenez-Gonzalez et al. emphasized front-end metrics such as PMI as drivers of innovation and not simply just problem-solving.

A front-end analysis is also effective when evaluating the greenness of a particular process. In this context, it is generally harder to determine waste and analyze E factors than it is to determine efficiency in terms of total material input using PMI. Confusion over what constitutes waste has resulted in the reporting of vastly different PMIs and E Factors [24–26, 28, 81]. It should be stressed that the difference between these two metrics should always be 1, and that both metrics should always account for the amount of recovered material. In the context of driving a green chemistry agenda, Federsel described the contributions of PMI for the pharmaceutical industry in a recent review [82]. Other reviews have emphasized the mathematical relationships between green metrics in more complex scenarios [7, 83–85].

3.2.3 Biocatalysis and the Synthesis of Singulair

In 2010, Liang et al. described a new biocatalytic process for an intermediate to montelukast sodium (Singulair, Fig. 3.5). Singulair is a leukotriene receptor antagonist developed by Merck for controlling the symptoms of asthma and allergies [86].

In the original synthesis, the asymmetric reduction of a ketone intermediate was carried out using excess amounts of (−)-B-chlorodiisopinocampheylborane [(−)-DIP-Cl, Scheme 3.12]. Despite being a mild and selective reagent, the corrosivity, moisture sensitivity, tedious workup and large waste stream associated with the use of (−)-DIP-Cl made this process very inefficient. In contrast, the biocatalytic alternative uses a novel ketoreductase enzyme (KRED) in conjunction with NAD (P) to selectively transfer a hydrogen atom from isopropanol to the same ketone

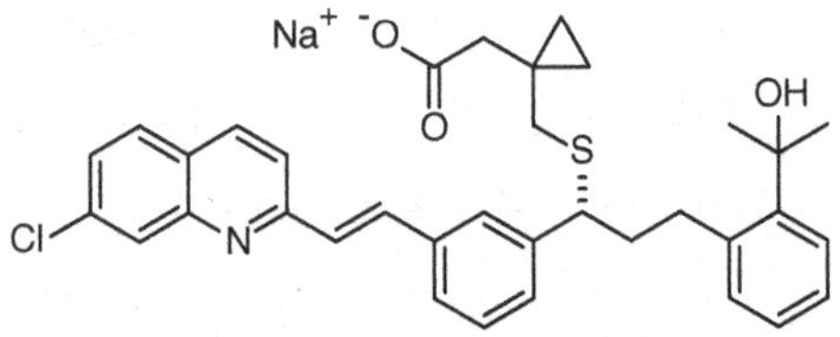

Fig. 3.5 Structure of montelukast sodium (Singulair)

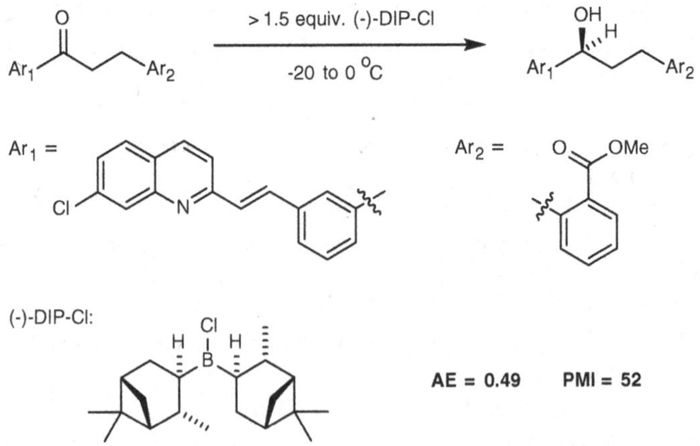

Scheme 3.12 (−)-DIP-Cl-mediated asymmetric reduction of a key Singulair precursor

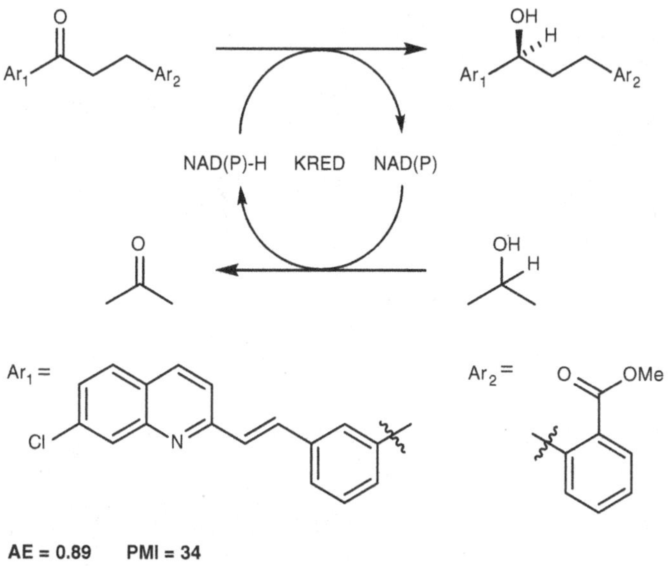

AE = 0.89   PMI = 34

Scheme 3.13 KRED-NAD(P)-catalyzed asymmetric reduction of a key Singulair precursor

(Scheme 3.13). This route has an atom economy of 89 % and a PMI of 34 as compared to 49 % and 52 for the original Merck synthesis. The PMI breakdown for the two processes is provided in Fig. 3.6.

The PMI of the KRED-NAD(P)-catalyzed synthesis of Singulair accounts for a production scale of 230 kg and represents a 30 % decrease in material input as compared to the (−)-DIP-Cl process. Furthermore, the KRED process uses fewer inorganic salts and 25 % less organic solvents, with those utilized being more environmentally friendly (e.g. methanol, ethyl acetate and isopropanol) [75]. Nevertheless, the KRED process still requires a large quantity of organic solvents to promote rate enhancement and product precipitation. Future research might therefore incorporate biocatalysis with alternative reaction media in order to achieve a greener synthesis. Similar studies using PMI for determining process greenness have been conducted on biocatalytic routes [87] as well as industrial syntheses carried out on a commercial scale [88].

3.2.4 Future Trends and the Changing Industrial Landscape

Global green metrics such as process mass intensity are preferred by the chemical industry because of their simple interpretation and application to real-world examples. Concerning a total synthesis, PMI is one of the easiest and fastest metrics to determine [89]. In addition, established chemistry journals such as *Organic Process Research and Development* have started to require that submitted articles

Fig. 3.6 Process mass intensity for the DIP-Cl and KRED-NAD(P)-catalyzed preparation of Singulair. Adapted with permission from [86]. Copyright 2010 American Chemical Society

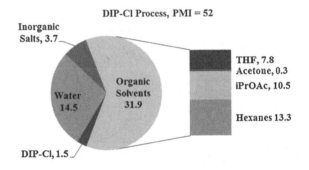

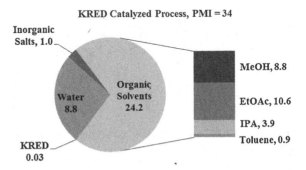

report green metrics calculations [90]. This has generated interest in establishing PMI target goals at pharmaceutical companies [91, 92], as well as the development of PMI tools to incorporate a life-cycle analysis for high-value targets such as pharmaceuticals [93]. Technical research focused on minimizing the environmental impact of syntheses through novel solvent recovery methods has also made significant progress [94].

Nevertheless, the PMI and E factor metrics are approaches for measuring process efficiency based on mass. As a result, they cannot be used to measure toxicity, operational cost, energy expenditure, and general safety associated with the use of hazardous substances. Although methods which calculate these parameters provide a bigger picture of a process, they also require chemists to sacrifice some degree of objectivity when deciding which values to assign to which parameters. A pertinent question is therefore: how does one arrive at a simple method whilst maintaining a respectable level of objectivity? Perhaps a more comprehensive approach that integrates both quantitative and qualitative criteria is the solution.

References

1. Sheldon RA (1992) Organic synthesis—past, present and future. Chem Ind 903–906
2. Sheldon RA (1997) Catalysis and pollution prevention. Chem Ind 12–15
3. Sheldon RA (2007) The E Factor: fifteen years on. Green Chem 1273–1283. doi:10.1039/b713736m

4. Sheldon RA (2008) E factors, green chemistry and catalysis: an odyssey. Chem Commun 29:3352–3365. doi:10.1039/b803584a
5. Sheldon RA (2012) Fundamentals of green chemistry: efficiency in reaction design. Chem Soc Rev 41:1437–1451. doi:10.1039/c1cs15219j
6. Thayer AM (2007) Chemists are finding asymmetric synthesis increasingly handy for making pharmaceutical compounds at large scale. Chem Eng News 85:11–19
7. Andraos J (2005) Unification of reaction metrics for green chemistry: applications to reaction analysis. Org Process Res Dev 9:149–163. doi:10.1021/op049803n
8. Calvo-Flores FG (2009) Sustainable chemistry metrics. ChemSusChem 2:905–919. doi:10.1002/cssc.200900128
9. Yang F, Cao J (2012) Biosynthesis of phloroglucinol compounds in microorganisms—review. Appl Microbiol Biotechnol 93:487–495. doi:10.1007/s00253-011-3712-6
10. Cao J, Jiang X, Zhang R, Xian M (2011) Improved phloroglucinol production by metabolically engineered *Escherichia coli*. Appl Microbiol Biotechnol 91:1545–1552. doi:10.1007/s00253-011-3304-5
11. Rao G, Lee J-K, Zhao H (2013) Directed evolution of phloroglucinol synthase PhlD with increased stability for phloroglucinol production. Appl Microbiol Biotechnol 97:5861–5867. doi:10.1007/s00253-013-4713-4
12. Climent MJ, Corma A, Iborra S, Mifsud M, Velty A (2010) New one-pot multistep process with multifunctional catalysts: decreasing the E factor in the synthesis of fine chemicals. Green Chem 12:99–107. doi:10.1039/b919660a
13. Stacey JM, Dicks AP, Goodwin AA, Rush BM, Nigam M (2013) Green carbonyl condensation reactions demonstrating solvent and organocatalyst recyclability. J Chem Educ 90:1067–1070. doi:10.1021/ed300819r
14. Aktoudianakis E, Chan E, Edward AR, Jarosz I, Lee V, Mui L, Thatipamala SS, Dicks AP (2009) Comparing the traditional with the modern: a greener, solvent-free dihydropyrimidone synthesis. J Chem Educ 86:730–732. doi:10.1021/ed086p730
15. Andraos J, Dicks AP (2012) Green chemistry teaching in higher education: a review of effective practices. Chem Educ Res Pract 13:69–79. doi:10.1039/c1rp90065j
16. Van Arnum SD (2005) An approach towards teaching green chemistry fundamentals. J Chem Educ 82:1689–1692. doi:10.1021/ed082p1689
17. McKenzie LC, Huffman LM, Hutchison JE (2005) The evolution of a green chemistry laboratory experiment: greener brominations of stilbene. J Chem Educ 82:306–310. doi:10.1021/ed082p306
18. Eissen M (2012) Sustainable production of chemicals—an educational perspective. Chem Educ Res Pract 13:103–111. doi:10.1039/c2rp90002e
19. Blacker AJ, Williams MT (2011) Pharmaceutical process development: current chemical and engineering challenges. Royal Society of Chemistry, Cambridge, pp 251–252
20. Strappaveccia G, Lanari D, Gelman D, Pizzo F, Rosati O, Curini M, Vaccaro L (2013) Efficient synthesis of cyanohydrin trimethylsilyl ethers via 1,2-chemoselective cyanosilylation of carbonyls. Green Chem 15:199–204. doi:10.1039/c2gc36442e
21. Ballerini E, Crotti P, Frau I, Lanari D, Pizzo F, Vaccaro L (2013) A waste-minimized protocol for the preparation of 1,2-azido alcohols and 1,2-amino alcohols. Green Chem 15:2394–2400. doi:10.1039/c3gc40988k
22. Song JJ, Reeves JT, Fandrick DR, Tan Z, Yee NK, Senanayake CH (2008) Achieving synthetic efficiency through new method development. Green Chem Lett Rev 1:141–148. doi:10.1080/17518250802592360
23. Mayo DW, Pike RM, Forbes DC (2013) Microscale organic laboratory with multistep and multiscale syntheses, 6th edn. Wiley, Hoboken, pp 421–427
24. Comerford JW, Clark JH, Macquarrie DJ, Breeden SW (2009) Clean, reusable and low cost heterogeneous catalyst for amide synthesis. Chem Commun 2562–2564. doi:10.1039/b901581g

25. Chaudhari PS, Salim SD, Sawant RV, Akamanchi K (2010) Sulfated tungstate: a new solid heterogeneous catalyst for amide synthesis. Green Chem 12:1707–1710. doi:10.1039/c0gc00053a
26. Ghosh S, Bhaumik A, Mondal J, Mallik A, Sengupta S, Mukhopadhyay C (2012) Direct amide bond formation from carboxylic acids and amines using activated alumina balls as a new, convenient, clean, reusable and low cost heterogeneous catalyst. Green Chem 14:3220–3229. doi:10.1039/c2gc35880h
27. Wang Z (2009) Comprehensive organic name reactions and reagents. Wiley, Hoboken, pp 2399–2404
28. Katkar VK, Chaudhari PS, Akamanchi KG (2011) Sulfated tungstate: an efficient catalyst for the Ritter reaction. Green Chem 13:835–838. doi:10.1039/c0gc00759e
29. Ichihashi H, Kitamura M (2002) Some aspects of the vapor phase Beckmann rearrangement for the production of ε-caprolactam over high silica MFI zeolites. Catal Today 73:23–28. doi:10.1016/S0920-5861(01)00514-4
30. Ichihashi H, Sato H (2001) The development of new heterogeneous catalytic processes for the production of ε-caprolactam. Appl Catal A 221:359–366. doi:10.1016/S0926-860X(01)00887-0
31. Sheldon RA, Dakka J (1994) Heterogeneous catalytic oxidations in the manufacture of fine chemicals. Catal Today 19:215–246. doi:10.1016/0920-5861(94)80186-X
32. Brown SH (2009) Zeolites in catalysis. In: Crabtree RH (ed) Handbook of green chemistry volume 2: heterogeneous catalysis. Wiley-VCH Verlag GmbH & Co. KGaA, Weinheim
33. Bellussi G, Perego C (2000) Industrial catalytic aspects of the synthesis of monomers for nylon production. CATTECH 4:4–16. doi:10.1023/A:1011905009608
34. Heveling J (2012) Heterogeneous catalytic chemistry by example of industrial applications. J Chem Educ 89:1530–1536. doi:10.1021/ed200816g
35. Sheldon RA (2005) Green solvents for sustainable organic synthesis: state of the art. Green Chem 7:267–278. doi:10.1039/b418069k
36. Papadogianakis G, Maat L, Sheldon RA (1997) Catalytic conversions in water. Part 5: Carbonylation of 1-(4-isobutylphenyl)ethanol to ibuprofen catalysed by water-soluble palladium–phosphine complexes in a two-phase system. J Chem Technol Biotechnol 70:83–91. doi:10.1002/(SICI)1097-4660(199709)70:1<83:AID-JCTB679>3.0.CO;2-7
37. Hintermair U, Francio G, Leitner W (2011) Continuous flow organometallic catalysis: new wind in old sails. Chem Commun 47:3691–3701. doi:10.1039/c0cc04958a
38. Pavia C, Ballerini E, Bivona LA, Giacalone F, Aprile C, Vaccaro L, Gruttadauria M (2013) Palladium supported on cross-linked imidazolium network on silica as highly sustainable catalysts for the Suzuki reaction under flow conditions. Adv Synth Catal 355:2007–2018. doi:10.1002/adsc.201300215
39. Cornils B, Herrmann WA, Horvath IT, Leitner W, Mecking S, Oliver-Bourbigou H, Vogt D (2005) Multiphase homogeneous catalysis, vol 1. Wiley-VCH Verlag GmbH & Co. KgaA, Weinheim
40. Sheldon RA (2000) Atom efficiency and catalysis in organic synthesis. Pure Appl Chem 72:1233–1246. doi:10.1351/pac200072071233
41. Clark JH (2002) Solid acids for green chemistry. Acc Chem Res 35:791–797. doi:10.1021/ar010072a
42. Sheldon RA (1997) Catalysis: the key to waste minimization. J Chem Technol Biotechnol 68:381–388. doi:10.1002/(SICI)1097-4660(199704)68:4<381:AID-JCTB620>3.0.CO;2-3
43. Bonollo S, Lanari D, Longo JM, Vaccaro L (2012) E-factor minimized protocols for the polystyryl-BEMP catalyzed conjugate additions of various nucleophiles to α, β-unsaturated carbonyl compounds. Green Chem 14:164–169. doi:10.1039/c1gc16088e
44. Dehaudt J, Husson J, Guyard L (2011) A more efficient synthesis of 4,4′,4″-tricarboxy-2,2′:6′,2″-terpyridine. Green Chem 13:3337–3340. doi:10.1039/c1gc15808b
45. Ciriminna R, Pagliaro M (2013) Green chemistry in the fine chemicals and pharmaceutical industries. Org Process Res Dev 17:1479–1484. doi:10.1021/op400258a
46. Sheldon RA, Arends IWCE, Hanefeld U (2007) Green chemistry and catalysis. Wiley-VCH Verlag, Weinheim, pp 29–34

47. Sheldon RA (2004) Green chemistry and catalysis for sustainable organic synthesis, Lecture given at University Pierre et Marie Curie, Paris, May 12, 2004. http://www.ed406.upmc.fr/cours/shaldon.pdf. Accessed 3 Feb 2014

48. Dunn PJ (2012) The importance of green chemistry in process research and development. Chem Soc Rev 41:1452–1461. doi:10.1039/c1cs15041c

49. Dunn PJ, Hettenbach K, Kelleher P, Martinez CA (2010) The development of a green, energy efficient, chemoenzymatic manufacturing process for pregabalin. In: Dunn PJ, Wells AS, Williams MT (eds) Green chemistry in the pharmaceutical industry. Wiley-VCH Verlag GmbH & Co, KGaA, Weinheim

50. Leuchs S, Na'amnieh S, Greiner L (2013) Enantioselective reduction of sparingly water-soluble ketones: continuous process and recycle of the aqueous buffer system. Green Chem 15:167–176. doi:10.1039/c2gc36558h

51. Wenda S, Illner S, Mell A, Kragl U (2011) Industrial biotechnology—the future of green chemistry? Green Chem 13:3007–3047. doi:10.1039/c1gc15579b

52. Constable DJC, Jimenez-Gonzalez C, Henderson RK (2007) Perspective on solvent use in the pharmaceutical industry. Org Process Res Dev 11:133–137. doi:10.1021/op060170h

53. Jimenez-Gonzalez C, Ponder CS, Broxterman QB, Manley JB (2011) Using the right green yardstick: why process mass intensity is used in the pharmaceutical industry to drive more sustainable processes. Org Process Res Dev 15:912–917. doi:10.1021/op200097d

54. Sauer ELO (2012) Organic reactions under aqueous conditions. In: Dicks AP (ed) Green organic chemistry in lecture and laboratory. CRC Press, Taylor and Francis Group, Boca Raton

55. Li CJ (2010) Handbook of green chemistry volume 5, green solvents: reactions in water. Wiley-VCH Verlag GmbH & Co. KGaA, Weinheim, pp 1–3, 151, 207–210, 215, 363–408

56. Cornils B, Herrmann WA (2004) Aqueous-phase organometallic catalysis, 2nd edn. Wiley-VCH Verlag GmbH & Co, KGaA, Weinheim

57. Isley NA, Gallou F, Lipshutz BH (2013) Transforming Suzuki—Miyaura cross-couplings of MIDA boronates into a green technology: no organic solvents. J Am Chem Soc 135:17707–17710. doi:10.1021/ja409663q

58. Leitner W, Jessop PG (2010) Handbook of green chemistry volume 4, green solvents: supercritical solvents. Wiley-VCH Verlag GmbH & Co. KGaA, Weinheim, pp 1–6, 77–79, 189–241

59. Kerton FM, Marriott R (2013) Alternative solvents for green chemistry, 2nd edn. RSC Publishing, Cambridge

60. Munshi P, Bhaduri S (2009) Supercritical CO_2: a twenty-first century solvent for the chemical industry. Curr Sci 97:63–72

61. Krämer A, Sabine M, Vogel H (1999) Hydrolysis of nitriles in supercritical water. Chem Eng Technol 22:494–500. doi:10.1002/(SICI)1521-4125(199906)22:6<494:AID-CEAT494>3.0.CO;2-U

62. Ranke J, Stolte S, Störmann R, Arning J, Jastorff B (2007) Design of sustainable chemical products—the example of ionic liquids. Chem Rev 107:2183–2206. doi:10.1021/cr050942s

63. Coleman D, Gathergood N (2010) Biodegradation studies of ionic liquids. Chem Soc Rev 39:600–637. doi:10.1039/b817717c

64. Wasserscheid P, Stark A (2010) Handbook of green chemistry volume 6, green solvents: ionic liquids. Wiley-VCH Verlag GmbH & Co. KGaA, Weinheim, pp 3–38

65. Kohlmann C, Leuchs S, Greiner L, Leitner W (2011) Continuous biocatalytic synthesis of (R)-2-octanol with integrated product separation. Green Chem 13:1430–1436. doi:10.1039/c0gc00790k

66. Constable DJC, Curzons AD, Cunningham VL (2002) Metrics to "green" chemistry—which are the best? Green Chem 4:521–527. doi:10.1039/b206169b

67. Martins MAP, Frizzo CP, Moreira DN, Buriol L, Machado P (2009) Solvent-free heterocyclic synthesis. Chem Rev 109:4140–4182. doi:10.1021/cr9001098

68. Dunn PJ, Galvin S, Hettenbach K (2004) The development of an environmentally benign synthesis of sildenafil citrate (Viagra™) and its assessment by green chemistry metrics. Green Chem 6:43–48. doi:10.1039/b312329d
69. Taber GP, Pfisterer DM, Colberg JC (2004) A new and simplified process for preparing N-[4-(3,4-dichlorophenyl)-3,4-dihydro-1(2H)-naphthalenylidene]methanamine and a telescoped process for the synthesis of (1S-cis)-4-(3,4-dichlorophenol)-1,2,3,4-tetrahydro-N-methyl-1-naphthalenamine mandelate: key intermediates in the synthesis of sertraline hydrochloride. Org Process Res Dev 8:385–388. doi:10.1021/op0341465
70. Rocha-Martin J, Velasco-Lozano S, Guisan JM, Lopez-Gallego F (2014) Oxidation of phenolic compounds catalyzed by immobilized multi-enzyme systems with integrated hydrogen peroxide production. Green Chem 16:303–311. doi:10.1039/c3gc41456f
71. Sheykhan M, Ranjbar ZR, Morsali A, Heydari A (2012) Minimisation of E-Factor in the synthesis of N-hydroxylamines: the role of silver(I)-based coordination polymers. Green Chem 14:1971–1978. doi:10.1039/c2gc35076a
72. Gruttadauria M, Giacalone F, Noto R (2013) "Release and catch" catalytic systems. Green Chem 15:2608–2618. doi:10.1039/c3gc41132j
73. Rundquist EM, Pink CJ, Livingston AG (2012) Organic solvent nanofiltration: a potential alternative to distillation for solvent recovery from crystallisation mother liquors. Green Chem 14:2197–2205. doi:10.1039/c2gc35216h
74. Content S, Dupont T, Fedou NM, Smith JD, Twiddle SJR (2013) Optimization of the manufacturing route to PF-610355 (1): synthesis of intermediate 5. Org Process Res Dev 17:193–201. doi:10.1021/op300341n
75. Prat D, Pardigon O, Flemming H-W, Letestu S, Ducandas V, Isnard P, Guntrum E, Senac T, Ruisseau S, Cruciani P, Hosek P (2013) Sanofi's solvent selection guide: a step toward more sustainable processes. Org Process Res Dev 17:1517–1525. doi:10.1021/op4002565
76. Trost BM (1991) The atom economy—a search for synthetic efficiency. Science 254:1471–1477. doi:10.1126/science.1962206
77. Heinzle E, Weirich D, Brogli F, Hoffmann VH, Koller G, Verduyn MA, Hungerbuhler K (1998) Ecological and economic objective functions for screening in integrated development of fine chemical processes. 1. flexible and expandable framework using indices. Ind Eng Chem Res 37:3395–3407. doi:10.1021/ie9708539
78. Eissen M, Metzger JO (2002) Environmental performance metrics for daily use in synthetic chemistry. Chem Eur J 8:3580–3585. doi:10.1002/1521-3765(20020816)8:16<3580:AID-CHEM3580>3.0.CO;2-J
79. Watson WJW (2012) How do the fine chemical, pharmaceutical, and related industries approach green chemistry and sustainability? Green Chem 14:251–259. doi:10.1039/c1gc15904f
80. Jimenez-Gonzalez C, Constable DJC, Ponder CS (2012) Evaluating the "Greenness" of chemical processes and products in the pharmaceutical industry—a green metrics primer. Chem Soc Rev 41:1485–1498. doi:10.1039/c1cs15215g
81. Das VK, Borah M, Thakur AJ (2013) Piper-betle-shaped nano-S-catalyzed synthesis of 1-amidoalkyl-2-naphthols under solvent-free reaction condition: a greener "nanoparticle-catalyzed organic synthesis enhancement" approach. J Org Chem 78:3361–3366. doi:10.1021/jo302682k
82. Federsel H-J (2013) En route to full implementation: driving the green chemistry agenda in the pharmaceutical industry. Green Chem 15:3105–3115. doi:10.1039/c3gc41629a
83. Auge J (2008) A new rationale of reaction metrics for green chemistry. Mathematical expression of the environmental impact factor of chemical processes. Green Chem 10:225–231. doi:10.1039/b711274b
84. Auge J, Scherrmann M-C (2012) Determination of the global material economy (GME) of synthesis sequences—a green chemistry metric to evaluate the greenness of products. New J Chem 36:1091–1098. doi:10.1039/c2nj20998e
85. Eissen M, Mazur R, Quebbbemann H-G, Pennemann K-H (2004) Atom economy and yield of synthesis sequences. Helv Chim Acta 87:524–535. doi:10.1002/hlca.200490050

86. Liang J, Lalonde J, Borup B, Mitchell V, Mundorff E, Trinh N, Kochrekar DA, Cherat RN, Pai GG (2010) Development of a biocatalytic process as an alternative to the (−)-DIP-Cl-mediated asymmetric reduction of a key intermediate of montelukast. Org Process Res Dev 14:193–198. doi:10.1021/op900272d

87. Gallou F, Seeger-Weibel M, Chassagne P (2013) Development of a robust and sustainable process for nucleoside formation. Org Process Res Dev 17:390–396. doi:10.1021/op300335d

88. Tian J, Shi H, Li X, Yin Y, Chen L (2012) Coupling mass balance analysis and multi-criteria ranking to assess the commercial-scale synthetic alternatives: a case study on glyphosate. Green Chem 14:1990–2000. doi:10.1039/c2gc35349k

89. Turgis R, Billault I, Acherar S, Auge J, Scherrmann M-C (2013) Total synthesis of high loading capacity PEG-based supports: evaluation and improvement of the process by use of ultrafiltration and PEG as a solvent. Green Chem 15:1016–1029. doi:10.1039/c3gc37097f

90. Laird T (2013) Green chemistry is good process chemistry. Org Process Res Dev 16:1–2. doi:10.1021/op200366y

91. Leahy DK, Tucker JL, Mergelsberg I, Dunn PJ, Kopach ME, Purohit VC (2013) Seven important elements for an effective green chemistry program: an IQ consortium perspective. Org Process Res Dev 17:1099–1109. doi:10.1021/op400192h

92. Kjell DP, Watson IA, Wolfe CN, Spitler JT (2013) Complexity-based metric for process mass intensity in the pharmaceutical industry. Org Process Res Dev 17:169–174. doi:10.1021/op3002917

93. Jimenez-Gonzalez C, Ollech C, Pyrz W, Hughes D, Broxterman QB, Bhathela N (2013) Expanding the boundaries: developing a streamlined tool for eco-footprinting of pharmaceuticals. Org Process Res Dev 17:239–246. doi:10.1021/op3003079

94. Kim JF, Szekely G, Valtcheva IB, Livingston AG (2014) Increasing the sustainability of membrane processes through cascade approach and solvent recovery—pharmaceutical purification case study. Green Chem 16:133–145. doi:10.1039/c3gc41402g

95. Dicks AP, Batey RA (2013) ConfChem conference on educating the next generation: green and sustainable chemistry—greening the organic curriculum: development of an undergraduate catalytic chemistry course. J Chem Educ 90:519–520. doi:10.1021/ed2004998

Chapter 4
Selected Qualitative Green Metrics

Abstract Qualitative green metrics such as the laboratory EcoScale, its modified industrial form and other computational tools are discussed. The first half of the chapter provides the reader with an appreciation of point-based categorical analysis as a means of assessing process greenness. Using calculated values for two benzodiazepine preparations, the EcoScale approach is explained in terms of its virtues and limitations. Simplicity and flexibility are highlighted as key advantages which make the EcoScale particularly effective in evaluating the nature of process operations and materials. Discussion then shifts to a recently proposed modified version of the EcoScale which addresses several drawbacks inherent in the original method. Using a newly defined system based on rewards rather than penalties, the modified EcoScale is shown to be effective at assessing industrial processes. The second portion of the chapter discusses two additional qualitative methods which are based on computational analysis. These approaches are the environmental assessment tool for organic syntheses (EATOS) and the radial polygon approach proposed by Andraos. To highlight their illustrative power and comprehensive scope, a recent article comparing four routes to a cyclic carbamate product is considered.

Keywords EcoScale · Modified EcoScale · EATOS · Radial pentagon · Environmental index · Benign index · Radial hexagon · Benzodiazepine synthesis · Cyclic carbamate

4.1 The EcoScale

In 2006, Van Aken et al. introduced the EcoScale as a semi-quantitative post-synthesis metric for analyzing specific organic reactions conducted in the laboratory [1]. As a score system based on penalties, the EcoScale assigns deductions to undesirable aspects of chemical preparations. Within the categories of "yield, cost, safety, conditions, and ease of workup and purification" [1], the overall greenness of a chemical transformation is evaluated based on a total score out of 100 points, where 100 describes ideal green reaction conditions. Since categories and deductions are more or less arbitrarily assigned, the EcoScale approach is easily modified to better represent the features of an experiment. In addition to its flexibility, the EcoScale also encompasses the nature of the materials and methods utilized. This feature allows scientists to assess a level of chemical complexity which the green metrics outlined in previous chapters cannot account for. In recent years, research on the EcoScale has focused on defining a similar method capable of evaluating industrial processes [2].

4.1.1 The Penalty System: Virtues and Drawbacks

The EcoScale is divided into six categories which establish penalty points according to the possible undesirable features of a laboratory experiment (see Table 1 in [1]). The six categories chosen by Van Aken et al. include product yield, price of reaction components, safety, technical setup, temperature/time, and workup/purification. In the product yield category, the penalty assigned is calculated by subtracting the isolated (pure) product yield from 100 and dividing the result in half. For example, an organic preparation with a final product yield of 70 % results in a deduction of 15 points. This calculation severely punishes lower isolated yields: indeed, when compared to other categories, it represents the largest penalty. To justify this point, the authors argued that low yields characterize an inefficient use of resources. Additionally, they noted that an increase in by-product formation often accompanies complicated work-up and purification steps.

Another aspect covered by the EcoScale pertains to the cost of reaction components. To account for expensive materials, deductions are made according to how much material is required to produce 10 mmol of the final product. For example, if it costs less than $10 of substance A to produce 10 mmol of the final product, the substance is considered inexpensive and is assigned no penalty. It is important to recognize that material cost depends on resource availability and technology, which means that penalties based on cost are not necessarily fixed across time. The authors reconcile this drawback by explaining that the choice of organic preparation and starting materials also depends on supply and demand. Moreover, cost-based penalties are likely to promote the development of simple procedures which feature an optimal use of reaction components (e.g. solvent-free techniques).

The third category of the EcoScale evaluates the safety of the chemicals used in an experiment. In order to simplify the calculation, the authors decided to have deductions based on the MSDS hazard warning symbol of each reaction component. Although it is possible to incorporate more complex evaluations, the virtue of the safety category is that it emphasizes the need to use safer chemicals in synthesis. A recent article describing an environmentally-benign peptide synthesis (Sect. 4.1.2) highlights research in this area [3].

The next EcoScale category analyzes the technical setup required for a chemical preparation. A setup consisting of a regular flask, a reflux condenser and a stir bar is awarded no penalty. Any extra equipment needed for the experiment receives a penalty corresponding to how much energy is required for it to operate.

In the fifth category, the duration and temperature under which an experiment is performed is considered. Since changes above or below ambient conditions require energy, the purpose of this penalty is to account for energy consumption due to changes in temperature. In this context, the authors noted that cooling is more energy-intensive than heating since conventional cooling (e.g. via an ice bath) provides access to fixed temperatures, whereas with heating one can easily access a continuous range of temperatures.

The final penalty category in the EcoScale accounts for the workup and purification step of a chemical procedure. Since this step can be lengthy and tedious, the authors designed penalties according to the principle of "time [required] to obtain the end product in a purity of over 98 %" [1]. As a result, all purification steps are judged based on two criteria: execution time and convenience. For example, cooling to room temperature is ranked lower than liquid-liquid extraction. The largest penalty in this category is awarded to classical chromatography, which is an often lengthy, tedious and very wasteful technique.

It is therefore clear that the EcoScale metric is a comprehensive and flexible analysis tool which promotes many aspects of green chemistry. In relation to green metrics discussed in previous chapters, the EcoScale is both clearly defined and simple to determine. The fact that one can evaluate otherwise complex features of experiments such as the nature of chemicals used, energy consumption, equipment, and practical techniques constitutes an advantage which other metrics do not have. Despite these virtues, the EcoScale approach is not without certain drawbacks, the most obvious of which surrounds the arbitrary assignment of categories and penalties. In certain respects, it is possible to argue that the EcoScale fails to consider essential aspects of process greenness such as the amount of solvent used, safety of solvents used, and the amount of generated waste including its environmental, safety and societal implications. This is why the EcoScale has not been significantly adopted by the chemical industry.

Although one can easily add new categories and penalties to address these concerns, a greater number of assessment criteria would further complicate the EcoScale making it more difficult for chemists to compare procedures with similar scores. The large emphasis on product yield is also problematic when applying the EcoScale in an educational setting since students are discouraged from making meaningful comparisons to literature procedures. Considering that students are often not capable of reproducing the high product yields typically reported in the primary literature, certain green aspects of their experiments are inevitably mitigated by low yields. Nevertheless, many of these issues can be resolved by making modifications to the EcoScale. Much of this work requires careful balancing acts with the aim of maintaining clarity and simplicity whilst developing ways to better account for process greenness.

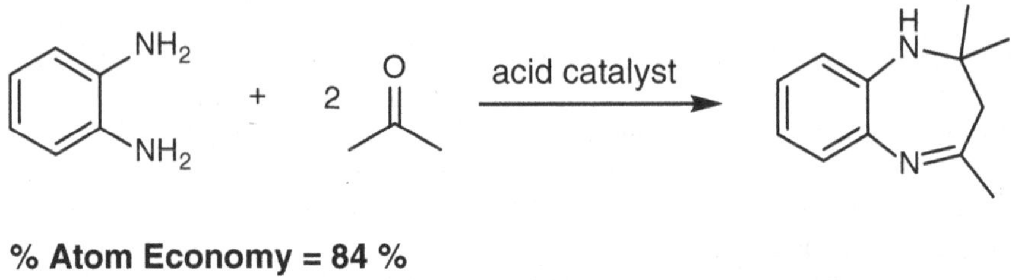

% **Atom Economy = 84 %**

Scheme 4.1 Acid-catalyzed condensation of 1,2-phenylenediamine with acetone to form a benzodiazepine product

4.1.2 Application in Education and Academia

It is clear that the EcoScale approach denotes a powerful idea in the context of green metrics. Although certain aspects of it are subjective and not necessarily representative of greenness, its comprehensive nature provides chemistry students with an opportunity to compare different preparations of a certain product from the point of view of a new metric. In the past 2 years, over 70 students at the University of Toronto have used the EcoScale approach to evaluate two preparations of a benzodiazepine product (Scheme 4.1) [4].

The two methods featured the Lewis acid catalysts zirconyl(IV) chloride [5] and sulfamic acid [6] and provided average student yields of 70 and 42 % respectively [4]. Following procedures adapted from the literature, students determined the EcoScale scores for both methods (Tables 4.1 and 4.2). Using these values, they were asked to compare their reactions with the literature methods and to draw conclusions about the strengths and weaknesses of the EcoScale. Lastly, students were required to suggest ways to improve the metric to better reflect the goals inherent in the Twelve Principles of Green Chemistry [4].

Applying the EcoScale in a classroom environment is also beneficial as one can easily calculate scores using a freely-accessible online tool [7]. Since 2006, a number of academic researchers have used the EcoScale to evaluate new chemistry carried out on a laboratory scale [3, 8–11]. Among these, three articles highlight some of the strengths and weaknesses of the EcoScale.

In 2010, Gaber et al. described a new enzymatic solvent-free production of *N*-alkanoyl-*N*-methylglucamide surfactants [10]. In order to assess environmental impact, the authors adopted several approaches which included the EcoScale. A comparison between new methods and literature procedures revealed acceptable to excellent EcoScale scores ranging from 63 to 96 (see Table 2 in [10]). Nevertheless, a closer inspection showed that the main factor contributing to the variation in EcoScale values was attributed to the penalty for isolated product yield where values ranged from 0.5 to 33. Overall, penalties in the other categories remained more or less the same for most other methods. It is interesting to note that the authors also used the E factor as part of their analysis. One could argue that in this case, because the experimental procedures are so similar, comparing process

Table 4.1 EcoScale calculation for the zirconyl (IV) chloride-catalyzed benzodiazepine synthesis

Parameter	Penalty points
1. Yield: 70 %	15
2. Price of Reaction Components	
1,2-phenylenediamine is expensive	3
3. Safety	
1,2-phenylenediamine (N), (T)	10
Zirconyl(IV) chloride (T)	5
Acetone (F)	5
4. Technical Setup	
Common setup	0
5. Temperature /Time	
Room temperature for <1 h.	0
6. Workup and Purification	
Adding solvent	0
Simple filtration	0
Removal of solvent with bp <150 °C	0
Liquid–liquid extraction	3
Total EcoScale score:	59

Table 4.2 EcoScale calculation for the sulfamic acid-catalyzed benzodiazepine synthesis

Parameter	Penalty points
1. Yield: 42 %	29
2. Price of reaction components	
1,2-phenylenediamine is expensive	3
3. Safety	
1,2-phenylenediamine (N), (T)	10
Sulfamic acid (T)	5
Acetone (F)	5
4. Technical setup	
Common setup	0
5. Temperature/time	
Room temperature for <1 h.	0
6. Workup and purification	
Adding solvent	0
Simple filtration	0
Removal of solvent with bp <150°C	0
Liquid–liquid extraction	3
Total EcoScale score:	45

features using the E factor is more advantageous. Note that the E factor also accounts for product yield (Sect. 3.1.2).

Problems surrounding the EcoScale's emphasis on yield were less pronounced in a recent article which described a ball mill-assisted environmentally benign peptide synthesis. In 2013, Bonnamour et al. applied heterogeneous ball-milling to catalyze the synthesis of dipeptides starting from Boc-protected α-amino acid anhydrides and succinimides reacting with α-amino acid ester salts [3]. In this case, the Eco-Scale was employed to compare the new method with two literature procedures. Although all procedures had product yields above 96 %, their EcoScale scores were calculated to be 84 for the novel synthesis and 69 and 58 respectively for the traditional approaches [3]. Upon further consideration, it turns out that the improvement in EcoScale value resulted from the use of much cheaper and safer reaction components. One can therefore conclude that when chemical preparations have similar product yields, differences in EcoScale scores tend to become more apparent. This is important because it allows for more meaningful assessments.

Lastly, the two routes to 4,4',4''-tricarboxy-2,2':6',2''-terpyridine (Sect. 3.1.4.2) highlight a significant use of the EcoScale in the context of a multistep synthesis [11]. Although the product yield was largely responsible for the stark difference in the EcoScale values, analyzing the penalty categories revealed certain green features of the homogeneous route. Although the penalties for safety were the same for both methods, the heterogeneous route used two substantially unsafe materials: potassium chromate (highly toxic and carcinogenic) and diethyl ether (highly flammable). In addition, the workup and purification steps of the heterogeneous route were also more complex. EcoScale scores for the homogeneous and heterogeneous routes were calculated to be 19.1 and 3.6 respectively. This, in addition to improvements in other metrics, legitimized the homogeneous catalysis approach over the traditional heterogeneous process.

4.1.3 The Modified Ecoscale: An Industrial Metric?

The EcoScale approach has proven valuable for comparing laboratory procedures, but its simplistic nature and limitations have largely prevented its adoption by the chemical industry. In 2012, Dach et al. described a modified EcoScale currently in use at Boehringer Ingelheim Pharmaceuticals [2]. This metric is similar to the original EcoScale in terms of awarding points to certain features of a process which are divided into eight categories. The difference is that instead of penalties, the modified EcoScale is based on rewards. For example, if the reaction time is less than 3 h, the process is awarded 10 points. If it is more, the number of points drops accordingly.

A similar breakdown of rewards (Table 2 in [2]) is provided for all other categories including product yield. In this instance, the modified EcoScale provides three different rewards for various ranges of yield. For example, a process with a product yield between 60 and 80 % receives 3 points whereas one with a yield between 80 and 95 % is awarded 7 points. In addition to reaction time and yield, the modified EcoScale also accounts for product quality (purity), workup/purification techniques, equipment, reaction temperature, raw materials and environmental health/safety information. Since reward-based systems do not have theoretical upward boundaries, it is easy to compare product routes in terms of tallying up accumulated strengths. Furthermore, this approach can motivate chemists to incorporate as many positive elements into a chemical synthesis as possible. The division of rewards between categories that are relevant in an industrial setting offers a solution to some of the limitations of the original EcoScale. It is interesting to note that Dach et al. extended their discussion by advocating for an integrated multi-metric-based system to evaluate processes. Consisting of eight criteria, relevant metrics are weighed according to the value and relevance of the information they produce. Despite the possible merits of this approach, it is clear that the flexibility inherent in the EcoScale allows for the rational development of more robust qualitative green metrics.

4.2 Other Qualitative Metrics

4.2.1 Environmental Assessment Tool for Organic Syntheses

As quantitative metrics became more widely adopted during the 1990s, chemists started developing ways of incorporating qualitative data into their environmental analyses. Regarding the E factor, the work of Hungerbuhler et al. [12, 13] helped define methods by which the environmental quotient factor Q [14] could be approximated. Using indices designed to estimate qualitative aspects such as raw material availability, environmental pollution and chemical toxicity, it became possible to evaluate the "environmental unfriendliness" characteristics of process materials. An overall score known as environmental index (EI) for both input and output materials could be determined by multiplying the E factor (E) and the environmental quotient factor (Q).

In 2002, Eissen and Metzger designed a freely-accessible software package known as environmental assessment tool for organic syntheses (EATOS) [15]. Using this tool, it was possible to calculate EI values for different procedures [16]. To make this possible, the user supplies the masses of all materials involved in the synthesis including specific information which can be easily accessed from MSDS databases. EATOS uses this information to determine mass flow data and to assign Q values from 1 to 10 to every material according to the potential environmental impact associated with the nature and identity of each substance. Overall input and output EI scores are presented in the form of histograms which identify the contribution and environmental impact of different classes of materials. Since their introduction, EATOS evaluations have appeared in numerous academic articles [17–20]. In some cases, the software was used simply to produce visual charts which illustrated the breakdown of input and output materials in the context of an E factor analysis [20]. In contrast to other quantitative metrics, EATOS has the advantage of accounting for the nature of process materials. This feature will be explored with an example described in the following section.

4.2.2 The Andraos Algorithm: Advancing Radial Polygons

At the end of Chap. 2, Andraos' integrated method for depicting efficiency parameters by means of radial pentagons [21] was briefly introduced. In a recent study, researchers in Italy applied this method in addition to EATOS to compare the performance of four different syntheses of 3-benzyl-1,3-oxazinan-2-one (Scheme 4.2) [22]. The purpose of the study was to determine which synthetic route was the greenest and to compare the effectiveness of both metrics in determining process greenness.

The first two routes (synthesis A and B) describe recent efforts at developing more environmentally-benign preparations of the final product while the last two procedures represent literature methods. It is important to note that the authors ignored workup and purification materials in their analysis. This was done because relevant details were not available in the literature. Using mass balance and appropriate MSDS information, the EATOS software produced histograms depicting both the input and output EI scores of all four syntheses. It was established that although synthesis D had the highest mass efficiency, synthesis A had

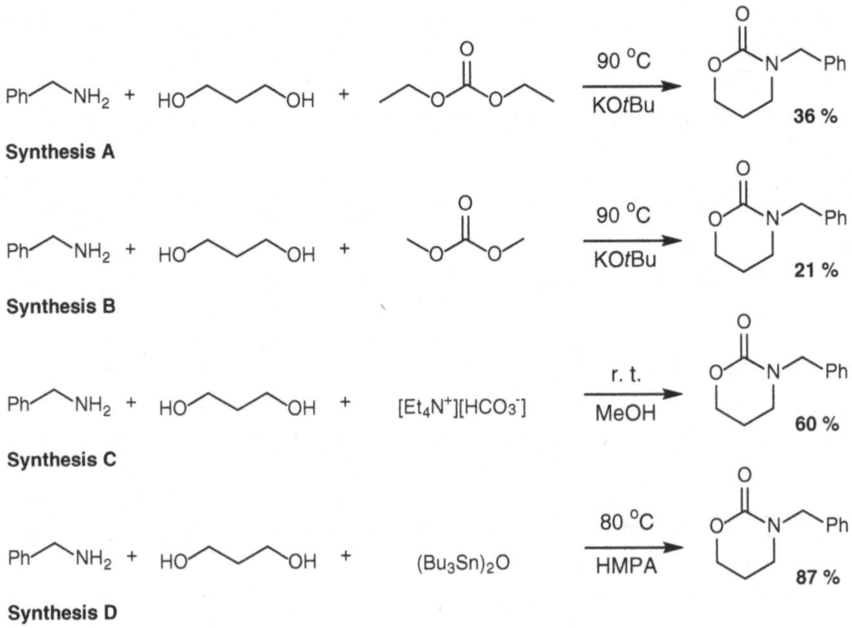

Scheme 4.2 Four synthetic routes for the production of 3-benzyl-1,3-oxazinan-2-one

the best score in terms of the environmental impact of input materials. The reason was that EATOS assigned much lower risk values to the substances used in synthesis A, in particular to the use of potassium *tert*-butoxide and diethyl carbonate. Synthesis D had a lower EI score due to the use of hexamethylphosphoramide (HMPA), a mildly toxic and carcinogenic polar aprotic solvent typically used to promote SN_2-type reactions.

In contrast, the Andraos spreadsheet algorithm showed that synthesis D was the greenest. This result was attributed to its higher mass efficiency. It is worth mentioning that according to EATOS, the environmental risk of this route was greater than for synthesis A, but not by very much. Additionally, although the product yield for synthesis D was double that of synthesis A, the atom economy of synthesis A was triple that of synthesis D. This discrepancy created an interesting result for the authors as they concluded that "a real green optimization for the synthesis of cyclic carbamates has not been achieved" since "the most material-efficient plan (synthesis D) [was] not also the most benign one (synthesis A)" [22]. Lastly, although applying both tools afforded greater insight into the green features of all four syntheses, the authors noted that EATOS provided an extra layer of qualitative evaluation as compared to Andraos' radial pentagon approach. To address this issue, in his most recent work, Andraos has proposed a sixth metric to be included in the radial polygon [23, 24]. This is called the benign index (BI) and is calculated using well-established hazard and life-cycle associated parameters (Chap. 5). With this new approach, the resulting radial hexagon accounts for the nature and environmental risk inherent in a synthesis.

4.2.3   Future Directions: What Does "Global" Really Mean?

A common theme of the metrics discussed so far in this chapter is that they all seek to provide a complete and "global" perspective on a chemical synthesis. This is done by incorporating qualitative information about environmental and toxicological risks of chemicals into an overall evaluation. Unfortunately, as one expands the reach of such evaluative methods beyond quantitative means, one inevitably introduces a certain degree of subjectivity into the analysis. Because of this, the importance of identifying and disclosing all calculations and details related to how certain methods are applied, the information that is included, and the limitations and assumptions made is once more highly emphasized. This practice was demonstrated in the few examples chosen to illustrate the metrics discussed in this chapter.

Although each of the qualitative metrics presented here have their own advantages and drawbacks, their inclusion into an environmental assessment can facilitate decision-making in both an academic and industrial setting. Nevertheless, it is possible (and important) to go even further so that aspects related to the entire lifecycle of a chemical process are considered and appropriately evaluated.

References

1. Van Aken K, Strekowski L, Patiny L (2006) EcoScale, a semi-quantitative tool to select an organic preparation based on economical and ecological parameters. Beilstein J Org Chem 2. doi:10.1186/1860-5397-2-3

2. Dach R, Song JJ, Roschangar F, Samstag W, Senanayake CH (2012) The eight criteria defining a good chemical manufacturing process. Org Process Res Dev 16:1697–1706. doi:10.1021/op300144g

3. Bonnamour J, Metro T-X, Martinez J, Lamaty F (2013) Environmentally benign peptide synthesis using liquid-assisted ball-milling: application to the synthesis of Leu-enkephalin. Green Chem 15:1116–1120. doi:10.1039/c3gc40302e

4. Dicks AP, Batey RA (2013) ConfChem conference on educating the next generation: green and sustainable chemistry—greening the organic curriculum: development of an undergraduate catalytic chemistry course. J Chem Educ 90:519–520. doi:10.1021/ed2004998

5. Nagawade R, Shinde DB (2006) Zirconyl(IV) chloride: a novel and efficient reagent for the rapid synthesis of 1,5-benzodiazepines under solvent-free conditions. Mendeleev Commun 16:113–115. doi:10.1070/MC2006v016n02ABEH002171

6. Fletcher JT, Boriraj G (2010) Benzodiazepine synthesis and rapid toxicity assay. J Chem Educ 87:631–633. doi:10.1021/ed100185n

7. The EcoScale, Fast and transparent evaluation of organic preparations (2006) http://www.ecoscale.org/. Accessed 22 May 2014

8. Escriba M, Eras J, Balcells M, Canela-Garayoa R (2013) H₃PO₄/metal halide induces a one-pot solvent-free esterification–halogenation of glycerol and diols. RSC Adv 3:8805–8810. doi:10.1039/c3ra41715h

9. Marcinkowska M, Rasala D, Puchala A, Galuszka A (2011) A comparison of green chemistry metrics for two methods of bromination and nitration of bis-pyrazolo[3,4-b;4′,3′-e]pyridines. Heterocycl Commun 17:191–196. doi:10.1515/HC.2011.038

10. Gaber Y, Tornvall U, Orellana-Coca C, Amin MA, Hatti-Kaul R (2010) Enzymatic synthesis of N-alkanoyl-N-methylglucamide surfactants: solvent-free production and environmental assessment. Green Chem 12:1817–1825. doi:10.1039/c004548a

11. Dehaudt J, Husson J, Guyard L (2011) A more efficient synthesis of 4,4′,4″-tricarboxy-2,2′:6′,2″-terpyridine. Green Chem 13:3337–3340. doi:10.1039/c1gc15808b

12. Heinzle E, Weirich D, Brogli F, Hoffmann VH, Koller G, Verduyn MA, Hungerbuhler K (1998) Ecological and economic objective functions for screening in integrated development of fine chemical processes. 1. Flexible and expandable framework using indices. Ind Eng Chem Res 37:3395–3407. doi:10.1021/ie9708539

13. Koller G, Fischer U, Hungerbuhler K (2000) Assessing safety, health, and environmental impact early during process development. Ind Eng Chem Res 39:960–972. doi:10.1021/ie990669i

14. Sheldon RA (2012) Fundamentals of green chemistry: efficiency in reaction design. Chem Soc Rev 41:1437–1451. doi:10.1039/c1cs15219j

15. EATOS (Environmental assessment tool for organic syntheses) (2014) http://www.metzger.chemie.uni-oldenburg.de/eatos/english.htm. Accessed 24 May 2014

16. Eissen M, Metzger JO (2002) Environmental performance metrics for daily use in synthetic chemistry. Chem Eur J 8:3580–3585. doi:10.1002/1521-3765(20020816)8:16<3580:AID-CHEM3580>3.0.CO;2-J

17. Eissen M, Hungerbuhler K, Dirks S, Metzger J (2003) Mass efficiency as metric for the effectiveness of catalysts. Green Chem 5:G25–G27. doi:10.1039/b301753m

18. Corradi A, Leonelli C, Rizzuti A, Rosa R, Veronesi P, Grandi R, Baldassari S, Villa C (2007) New "green" approaches to the synthesis of pyrazole derivatives. Molecules 12:1482–1495. doi:10.3390/12071482

19. Ravelli D, Protti S, Neri P, Fagnoni M, Albini A (2011) Photochemical technologies assessed: the case of rose oxide. Green Chem 13:1876–1884. doi:10.1039/c0gc00507j

20. Dunn PJ, Galvin S, Hettenbach K (2004) The development of an environmentally benign synthesis of sildenafil citrate (Viagra™) and its assessment by green chemistry metrics. Green Chem 6:43–48. doi:10.1039/b312329d

21. Andraos J, Sayed M (2007) On the use of "green" metrics in the undergraduate organic chemistry lecture and lab to assess the mass efficiency of organic reactions. J Chem Educ 84:1004–1010. doi:10.1021/ed084p1004

22. Toniolo S, Arico F, Tundo P (2014) A comparative environmental assessment for the synthesis of 1,3-oxazin-2-one by metrics: greenness evaluation and blind spots. ACS Sustain Chem Eng 2:1056–1062. doi:10.1021/sc500070t

23. Andraos J (2012) Inclusion of environmental impact parameters in radial pentagon material efficiency metrics analysis: using benign indices as a step towards a complete assessment of "greenness" for chemical reactions and synthesis plans. Org Process Res Dev 16:1482–1506. doi:10.1021/op3001405

24. Andraos J (2013) Safety/hazard indices: completion of a unified suite of metrics for the assessment of "greenness" for chemical reactions and synthesis plans. Org Process Res Dev 17:175–192. doi:10.1021/op300352w

Chapter 5
An Introduction to Life Cycle Assessment

Abstract An introduction to the history of life cycle assessment (LCA) is provided as a segue into a more detailed presentation of the guidelines and principles of LCA. The term "life cycle" is defined and illustrated by means of a figure which is used to describe various system boundaries that a typical LCA can evaluate. Discussion then shifts to the four stages of the LCA process identified in the international standards for conducting a life cycle analysis. These principles are connected with the principles of green chemistry in order to highlight common goals related to both fields. The theoretical portion ends with a discussion of the virtues and limitations inherent in the currently accepted methodology. This section covers the notion of "burden shifting" as well as the complexity and arbitrary nature of several components of the LCA process. The use of software packages to simplify LCA is described in the context of an approach developed at GlaxoSmithKline known as Fast Life Cycle Assessment of Synthetic Chemistry (FLASC^{TM}). The advantages of this software are explored by revisiting the synthesis of 7-aminocephalosporanic acid. The chapter concludes with a novel approach to teaching LCA in the context of green metrics to upper-level undergraduate students.

Keywords Life cycle assessment · Inventory analysis · Impact assessment · Burden-shifting · Potential environmental impact · Cradle-to-grave · System boundary · FLASC · 7-Aminocephalosporanic acid · Streamlined software

5.1 History and the Journey Toward Standardization

The study of life cycle assessment (LCA) originated with the environmental movement of the 1970s. Growing public concern over resource availability and energy use inspired researchers at the Coca-Cola Company to undertake one of the very first LCA studies in 1969 [1]. An internal review was conducted to determine which beverage container had the lowest emissions to the environment and the least effect on the supply chain [2]. The results of this study laid the groundwork for the

© The Author(s) 2015
A.P. Dicks and A. Hent, *Green Chemistry Metrics*,
SpringerBriefs in Green Chemistry for Sustainability,
DOI 10.1007/978-3-319-10500-0_5

use of LCA in the United States. In the following years, the increase in life cycle inventory assessments (LCIAs) performed by commercial institutions caused a negative shift in public opinion regarding the inappropriate use of LCA [3]. Criticism directed at the lack of a standardized methodology led to the formation of private and governmental bodies charged with developing international LCA standards. Over the years, these collaborations [4–6] have generated more rigorous LCA guidelines and principles, helping to evolve the field from simple inventory analysis to a general impact assessment methodology. From 1999 to 2000, the publication of two reviews highlighted the benefits of LCA analysis in the context of green chemistry [7–9]. Since then, the study of LCA has grown to encompass new journals dedicated exclusively to research and development [10], software tools capable of performing faster assessments more easily [11], as well as academic articles with a focus on introducing LCA to university students and faculty members [12].

5.2 Life Cycle Assessment (LCA)

5.2.1 The Nuts and Bolts

Life cycle assessment (LCA) is an analysis tool capable of evaluating the environmental impact of products and processes across their entire life cycle. It is useful to think of the life cycle as a collection of all the technical steps that are needed for and caused by the existence of the product (Fig. 5.1). According to Lancaster, this consists of "raw material production, manufacture, distribution, use and disposal" including all intermittent "transportation steps" [13]. A life cycle which begins with raw material acquisition and ends with the product's ultimate disposal is called "cradle-to-grave".

Despite its comprehensive nature, cradle-to-grave LCA often requires large volumes of data making the approach very costly and time-consuming. Consequently, researchers have defined various system boundaries appropriate for different types of life cycle assessments (Table 5.1). The LCA process, according to current international standards and guidelines, is divided into a sequence of four steps: goal definition and scope, inventory analysis, impact assessment, and interpretation [3, 6, 13]. The first and most important step is essentially the planning phase. This is where the objectives of the LCA are clearly identified. It is also where one establishes what information is required, the level of specificity needed and the best means to organize and present final results.

An additional aspect pertains to properly defining the system boundary within which environmental impacts are evaluated. For example, if material recycling/reuse is an essential part of the product life cycle, a cradle-to-cradle approach is recommended. This would also be the case in a study designed to compare the environmental impact of different recycling options for the end-of-life disposal step of a

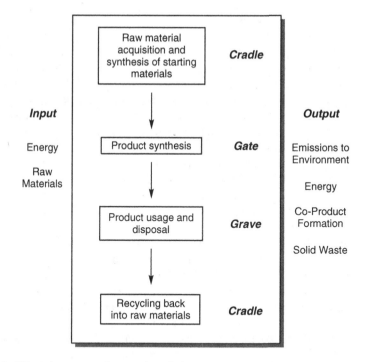

Fig. 5.1 Life cycle stages and system boundaries

Table 5.1 Possible ways to define the scope of an LCA

Type of LCA	Definition of system boundary
Cradle-to-grave	All life cycle stages which describe a process from extraction of raw materials to their return to the earth are considered. These stages include resource acquisition, product manufacture, use, and disposal, and all intermittent transportation steps
Cradle-to-gate	A cradle-to-gate analysis considers all stages from raw material production to the manufacture of the final product. Downstream (post-manufacturing) steps are assumed to remain stable for different processes. This analysis can be used to compare processes, usually at the same facility/company
Gate-to-gate	Gate-to-gate represents a partial life cycle which excludes the raw material acquisition stage. The scope of this analysis is generally confined to a single process at a single manufacturing facility. LCA results can be used for difference analysis or for developing cradle-to-gate and cradle-to-grave LCAs
Gate-to-grave	Gate-to-grave LCAs evaluate only life cycle stages which are found downstream of product manufacturing. These include product use, disposal and recycling
Cradle-to-cradle	A cradle-to-cradle life cycle is often referred to as a closed-loop system. This scenario occurs when the end-of-life disposal step for the product is a recycling process. One can use this approach to evaluate products which circulate in cycles of production, use, recovery and remanufacture (e.g. high-tech synthetics and mineral resources)

particular product. Moreover, just as one can evaluate a product and its manufacturing route(s), it is also possible to compare different products which have a similar function. Without proper definition of goals and boundaries, it is easy to exhaust one's available resources. A unifying principle one should strictly adhere to therefore is that of uncompromising simplicity. In other words, the scope of the study should be broad enough to include only the relevant areas where environmental burdens may be present and/or potentially avoided. For example, including all possible products of a high volume chemical within the LCA may produce irrelevant results at the price of significant time and capital expenditure [13]. The second stage of the LCA process is that of inventory analysis. During this phase, the scientist collects data to construct mass and energy flow charts which accurately describe the steps within the established system boundary. This data is generally quantitative in nature and often already known. To make the task simpler, process inputs, outputs and required amounts can be labelled directly onto a life cycle diagram.

When pertinent information is not available, one can turn to several alternatives such as approximation, literature searching, assumption, or redefinition of system boundaries. Whenever these options are employed, authors are encouraged to disclose the relevant details behind their decision-making process. With the number of yearly LCAs conducted steadily rising, the development of databases for LCA information on products and materials will eventually address the problem of data availability. The final step of inventory analysis concerns how collected data is to be presented. Although general consensus recommends constraining emissions to a certain medium (i.e. x amount of CO_2 to atmosphere) [3, 6], Lancaster explains that one can also cite data in terms of potential environmental impacts [13]. This type of reporting combines the information collected during the inventory analysis phase with impact categories identified in the impact assessment phase of the LCA process. A more in-depth discussion of these issues is available in other publications [3, 6].

The purpose of the third stage (impact assessment) is to help evaluate the potential effects of energy and material usage that correspond to the environmental emissions already identified. Numerical values are assigned within specific environmental impact categories selected to address the potential human, ecological and resource depletion effects relevant to a product's life cycle. Examples of these categories, among many others, include: aquatic toxicity, eutrophication potential, global warming potential, human toxicity potential, resource depletion and land use. For a detailed presentation along with actual calculations, the reader is referred to reference [6]. It is also important to consider the characterization (science-based conversion factors), normalization (ease of comparison) and weighing of potential impacts during this stage of the LCA [3]. These aspects have been discussed by others within an industrial [6] and a pedagogical [12] context.

The final stage of a life cycle assessment requires proper interpretation of the results obtained throughout the antecedent stages. The purpose of this step is to evaluate the merits and limitations of every aspect of the life cycle under analysis, including the analysis itself. Since LCA does not consider cost, technical performance or societal implication, it is not expected that individuals and companies can or should select products and processes based on LCA results alone. Instead, the results of an LCA should be thoroughly checked for completeness, consistency and sensitivity, and then incorporated into the wider decision-making process.

5.2.2 The Green Chemistry Connection

As the field of LCA became more established in the 1990s, scientists interested in green chemistry began to recognize the benefits of life cycle thinking and analysis. In 1999, Azapagic published a thorough review in *Chemical Engineering Journal* explaining the use of LCA in process selection, design and optimization [7]. Starting with the basic principles of LCA discussed within a historical context, the author analyzed various case studies of LCA applied to industrial processes. In terms of process selection, the article highlighted processes for the abatement of SO_2, NO_x and VOC (volatile organic compound). Of particular interest were the charts which clearly depict the environmental emissions of processes according to impact assessment categories. Similarly, the ways in which LCA can drive process optimization and design were also acknowledged.

This work contributed to a review written by Anastas and Lankey in 2000 connecting life cycle analysis with green chemistry [8]. In this review, the authors made the point that the practice of green chemistry can positively influence all aspects of the life cycle. This is because green chemistry has the ability to effect "changes in the hazard of a product at the most fundamental level, the molecular level" [8]. Translated, this statement reflects Principle 4 of green chemistry: designing safer chemicals. In conjunction with the other 11 Principles, it is possible to address life cycle impacts at the early design stage of the product or manufacturing process being considered. For example, exploring alternative feedstocks (Principle 7) can reduce the impacts for the raw material acquisition stage of the life cycle. Similarly, atom economy, catalysis and waste reduction can contribute to significantly improve the greenness of the manufacturing and processing stage of the life cycle. Designing biodegradable products (Principle 10) can improve the impact of the product's end-of-life disposal phase. Use of alternative reaction media, along with solvent recycling, can also reduce waste, toxicity and emissions to the environment. One can therefore conclude that the pursuit of green chemistry complements well with the study and development of LCA. Over the years, specific examples solidifying these ideas have appeared in the context of the LCA of metals [14], solvents [15, 16], industrial case studies [17, 18], and fundamental theory [19, 20].

5.2.3 Virtues and Limitations

An essential virtue of LCA is that the entire product life cycle is considered and investigated. The consequence of this is the limiting and sometimes complete prevention of the phenomenon known as "burden shifting". Burden shifting happens when attempts to reduce environmental burdens at one stage of the life cycle create impacts at other stages. For instance, a process with more environmentally friendly starting materials may produce greater life cycle impacts when considering how those starting materials are themselves manufactured. Lancaster has captured this point beautifully by outlining the manufacture of polycarbonates with and without phosgene [13]. Another example is the use of biofuels as an alternative energy source. Although proponents of the new technology advocate its renewability, LCA analysis reveals that issues associated with land and water usage do not support the claim of sustainability [21]. Since LCA covers environmental emissions in general, burden shifting between different environmental media is also addressed. When compared to other green metrics, LCA analysis is the only method which accounts for these effects. Although this feature does not eliminate the issue of "shades of green" which often arise when conducting environmental assessments, it certainly reduces the extent of the "shading effect".

Despite this benefit, LCA is also plagued by many limitations. By its very nature as a comprehensive tool capable of evaluating entire life cycles, LCA is a very resource intensive approach. Without a proper definition of its scope and goals, LCA can easily become very expensive. Furthermore, even though LCA data continues to be gathered for different products and materials, the availability of information is another potential limitation. A significant amount of method simplification is therefore required. A consequence of this is constraining LCA to linear models which do not account for changes in the economy, the environment, or society [6]. In addition, many environmental impacts are not contextualized in time and space and are often dependent on arbitrary unit definitions based on technical assumptions and value choices. The complexity of the process can also easily intimidate newcomers, thus creating difficulty when choosing to introduce the concept to chemistry and engineering students. Nevertheless, with greater transparency and further development, these limitations are slowly being understood and addressed. Some of these developments, including a novel pedagogical approach to teaching LCA in university courses, are discussed in Sect. 5.4.

5.3 Industrial Application: Revisiting the Synthesis of 7-Aminocephalosporanic Acid

In recent years, applications of LCA in the chemical industry largely depended upon use of simple, streamlined software tools [11]. One of these packages was developed at GlaxoSmithKline (GSK) and is called "Fast Life Cycle Assessment of Synthetic Chemistry" (FLASC^{TM}) [22]. FLASC is a software package which provides gate-to-gate life cycle analysis of chemical processes carried out at GSK. Because its base data set was developed using information obtained at GSK, FLASC is not accessible to the public. The core of this data consists of life cycle impact (LCI) information developed for 140 materials that were used in 22 well-established GSK processes. The methodology for collecting this data was described in 2000 by Jimenez-Gonzalez [10, 23]. A detailed LCI analysis for the manufacture of sertraline (Sect. 1.2.4) was also described in a PhD dissertation [24]. Using this information, FLASC is capable of addressing the raw material acquisition stage of the life cycle. Along with mass balance information divided among impact assessment categories, the software produces a process rating ranging from 1.0 to 5.0 [22]. A rating of 2.3 corresponds to the average life cycle environmental impact of 25 GSK routes developed between 1990 and 2000. Scores of 5.0 and 1.0 correspond to 12 and 300 % respectively of the total life cycle mass and energy associated with the average route.

Since its conception in 2007, FLASC analysis has been applied numerous times, most notably in comparing the chemical and enzymatic syntheses of 7-aminocephalosporanic acid (Sect. 2.2.2.3). Using FLASC, it has been shown that the enzymatic process has a score of 3.1 while the chemical process has a score of 2.7 [25]. In addition, the energy aspect of the chemical route was identified as particularly impactful. On the other hand, the biocatalysis route scored very well in the environment and safety categories. Together with the multi-metric analysis described in Chap. 2, the authors emphasized that the enzymatic process has many green features which, according to FLASC, do not simply push environmental burden to other parts of the life cycle. Compared to earlier research [26], this study highlights the progress made in developing simpler LCA tools for use in academic and industrial settings. The extent of this progress along with further examples has been described in several recent articles and reviews [27–32].

5.4 Future Directions: A Novel Approach to Teaching LCA and Green Metrics

One of the biggest challenges in introducing life cycle assessment to upper-level chemistry undergraduates has to do with the extremely complex nature of LCA. Introductory articles and publications over the years have only attempted to describe the practice of LCA, either in isolation [3, 4, 6, 13] or in the wider context of green metrics [33, 34]. Unfortunately, these approaches have not reproduced entire LCAs in terms of their particular stages with details on how each stage is resolved. Hence, lacking opportunities to observe and conduct an independent analysis, students are left to evaluate the results obtained by others. Consequently, students are somewhat inhibited from integrating the LCA process in their minds. Although there are few excellent sources where this is not the case [3, 24], these are generally rare and often written in language which is difficult for an introductory-level student to appreciate. In terms of industrial applications, it is often the case that full-scale LCA analysis is contracted to outside companies which produce only results without disclosing the details that make those results possible. To resolve this problem, one requires both the eagerness to work directly according to the principles of LCA and the foresight to establish clear methods that others can independently test and apply according to those principles.

Mercer et al. have recently described such an approach in the context of a multi-metric green chemistry exercise which asked students to evaluate the greenness of several industrial preparations of aniline [12]. To aid with this task, students were expected to calculate green metrics such as atom economy and E factor for all given processes. In addition, they also had to determine nine commonly-used LCA impact assessment metrics for all materials involved in the various syntheses. Since data required for LCI calculations was available in the literature, students had the ability to conduct real impact assessments for five different industrial processes. Using these results, the final task was to create a short presentation answering the question: which is the greenest synthesis?

Although this approach is somewhat complex and most likely appropriate for a fourth year undergraduate course, its ability to place leading-edge methodology in the hands of students interested in LCA make it very promising from a pedagogical standpoint. The opportunity to evaluate results in order to answer important questions may also be viewed as a highly motivational aspect for students who most likely have had no previous experience with the subject. Such a project would also provide educators with the opportunity to emphasize the values and limitations of green metrics and life cycle analysis. Using this approach it is possible to communicate the importance of discussing, presenting and criticizing all results and methods which are gathered and applied during the course of the project. If students can successfully adopt these practices, they will be empowered to develop more objective green metrics which will drive the practice of green chemistry for years to come.

References

1. Reduce (2012) http://www.coca-colacompany.com/stories/reduce. Accessed 7 May 2014
2. Guinee JB, Heijungs R, Huppes G, Zamagni A, Masoni P, Buonamici R, Ekvall T, Rydberg T (2011) Life cycle assessment: past, present and future. Environ Sci Tech 45:90–96. doi:10.1021/es101316v
3. Life cycle assessment: principles and practice (2006). http://nepis.epa.gov/Exe/ZyPDF.cgi/P1000L86.PDF?Dockey=P1000L86.PDF. Accessed 9 May 2014
4. Curran MA (1996) Environmental life-cycle assessment. McGraw-Hill, New York
5. Schaltegger S (1996) Life cycle assessment (LCA)—quo vadis?. Birkhauser Verlag, Basel
6. Guinee JB (2002) Handbook on life cycle assessment: operational guide to the ISO standards. Kluwer Academic Publishers, New York
7. Azapagic A (1999) Life cycle assessment and its application to process selection, design and optimisation. Chem Eng J 73:1–21. doi:10.1016/S1385-8947(99)00042-X
8. Anastas PT, Lankey RL (2000) Life cycle assessment and green chemistry: the yin and yang of industrial ecology. Green Chem 2:289–295. doi:10.1039/b005650m
9. Curran MA (2013) Life cycle assessment: a review of the methodology and its application to sustainability. Curr Opin Chem Eng 2:273–277. doi:10.1016/j.coche.2013.02.002
10. Gonzalez-Jimenez C, Curzons AD, Constable DJC, Cunningham VL (2004) Cradle-to-gate life cycle inventory and assessment of pharmaceutical compounds. Int J LCA 9:114–121. doi:10.1065/lca2003.11.141

11. Vervaeke M (2012) Life cycle assessment software for product and process sustainability analysis. J Chem Educ 89:884–890. doi:10.1021/ed200741b

12. Mercer SM, Andraos J, Jessop PG (2012) Choosing the greenest synthesis: a multivariate metric green chemistry exercise. J Chem Educ 89:215–220. doi:10.1021/ed200249v

13. Lancaster M (2010) Green chemistry: an introductory text, 2nd edn. RSC Paperbacks, Cambridge, pp 66–72

14. Lankey RL, Anastas PT (2002) Life-cycle approaches for assessing green chemistry technologies. Ind Eng Chem Res 41:4498–4502. doi:10.1021/ie0108191

15. Clark JH, Tavener SJ (2007) Alternative solvents: shades of green. Org Process Res Dev 11:149–155. doi:10.1021/op060160g

16. Raymond MJ, Slater CS, Savelski MJ (2010) LCA approach to the analysis of solvent waste issues in the pharmaceutical industry. Green Chem 12:1826–1834. doi:10.1039/c003666h

17. Burgess AA, Brennan DJ (2001) Application of life cycle assessment to chemical processes. Chem Eng Sci 56:2589–2604. doi:10.1016/S0009-2509(00)00511-X

18. Kralisch D (2009) Application of LCA in process development. In: Lapkin A, Constable DJC (eds) Green chemistry metrics: measuring and monitoring sustainable processes. Wiley-Blackwell, Chichester

19. Herrchen M, Werner K (2000) Use of the life-cycle assessment (LCA) toolbox for an environmental evaluation of production processes. Pure Appl Chem 72:1247–1252. doi:10. 1351/pac200072071247

20. Tukker A (2002) Life-cycle assessment and the precautionary principle. Environ Sci Technol 36:70A–75A. doi:10.1021/es022213p

21. Schebek L (2009) Life-cycle analysis of biobased products. In: Ulber R, Sell D, Hirth T (eds) Renewable raw materials. Wiley-VCH Verlag GmbH & Co, KGaA, Weinheim

22. Curzons AD, Jimenez-Gonzalez C, Duncan AL, Constable DJC, Cunningham VL (2007) Fast life cycle assessment of synthetic chemistry (FLASC^{TM}) tool. Int J LCA 12:272–280. doi:10. 1065/lca2007.03.315

23. Jimenez-Gonzalez C, Kim S, Overcash MR (2000) Methodology for developing gate-to-gate life cycle inventory information. Int J LCA 5:153–159. doi:10.1065/lca2000.02.017

24. Jimenez-Gonzalez C (2000) Life cycle assessment in pharmaceutical applications. Dissertation, North Carolina State University

25. Henderson RK, Jimenez-Gonzalez C, Preston C, Constable DJC, Woodley JM (2008) EHS & LCA assessment for 7-ACA synthesis, a case study for comparing biocatalytic & chemical synthesis. Ind Biotechnol 4:180–192. doi:10.1089/ind.2008.4.180

26. Domenech X, Ayllon JA, Peral J, Rieradevall J (2002) How green is a chemical reaction? Application of LCA to green chemistry. Environ Sci Technol 36:5517–5520. doi:10.1021/es020001m

27. Jimenez-Gonzalez C, Constable DJC, Ponder CS (2012) Evaluating the "Greenness" of chemical processes and products in the pharmaceutical industry—a green metrics primer. Chem Soc Rev 41:1485–1498. doi:10.1039/c1cs15215g

28. Dunn PJ (2013) Pharmaceutical green chemistry process changes—how long does it take to obtain regulatory approval? Green Chem 15:3099–3104. doi:10.1039/c3gc41376d

29. Huebschmann S, Kralisch D, Loewe H, Breuch D, Petersen JH, Dietrich T, Scholz R (2011) Decision support towards agile eco-design of microreaction processes by accompanying (simplified) life cycle assessment. Green Chem 13:1694–1707. doi:10.1039/c1gc15054e

30. Tufvesson LM, Tufvesson P, Woodley JM, Borjesson P (2013) Life cycle assessment in green chemistry: overview of key parameters and methodological concerns. Int J Life Cycle Assess 18:431–444. doi:10.1007/s11367-012-0500-1

31. Wang Q, Gursel IV, Shang M, Hessel V (2013) Life cycle assessment for the direct synthesis of adipic acid in microreactors and benchmarking to the commercial process. Chem Eng J 234:300–311. doi:10.1016/j.cej.2013.08.059

32. Shonnard D, Lindner A, Nguyen N, Ramachandran PA, Fichana D, Hesketh R, Slater CS, Engler R (2012) Green engineering: integration of green chemistry, pollution prevention, and

risk-based considerations. In: Kent JA (ed) Handbook of industrial chemistry and biotechnology, vol 1 and 2, 12th edn. Springer, New York

33. Constable DJC, Jimenez-Gonzalez CC (2012) Evaluating the greenness of synthesis. In: Li CJ (ed) Handbook of green chemistry volume 7: green synthesis. Wiley-VCH Verlag GmbH & Co. KGaA, Weinheim

34. Eissen M (2012) Sustainable production of chemicals—an educational perspective. Chem Educ Res Pract 13:103–111. doi:10.1039/c2rp90002e

Printed in the United States
By Bookmasters